MICROWAVE BANDPASS FILTERS FOR WIDEBAND COMMUNICATIONS

WILEY SERIES IN MICROWAVE AND OPTICAL ENGINEERING

KAI CHANG, Editor
Texas A&M University

A complete list of the titles in this series appears at the end of this volume.

MICROWAVE BANDPASS FILTERS FOR WIDEBAND COMMUNICATIONS

LEI ZHU
Nanyang Technological University, Singapore

SHENG SUN
The University of Hong Kong, Hong Kong

RUI LI
Institute of Microelectronics, Singapore

A JOHN WILEY & SONS, INC., PUBLICATION

Published by John Wiley & Sons, Inc., Hoboken, New Jersey

Published simultaneously in Canada

For general information on our other products and services or for technical support, please contact our Customer Care Department within the United States at (800) 762-2974, outside the United States at (317) 572-3993 or fax (317) 572-4002.

Wiley also publishes its books in a variety of electronic formats. Some content that appears in print may not be available in electronic formats. For more information about Wiley products, visit our web site at www.wiley.com.

Library of Congress Cataloging-in-Publication Data:
Zhu, Lei, 1963-
  Microwave bandpass filters for wideband communications / Lei Zhu, Sheng Sun, Rui Li.
      p. cm.
  Includes bibliographical references.
  ISBN 978-0-470-87661-9 (hardback).
  1. Broadband communication systems.   2. Ultra-wideband devices.   I. Sun, Sheng,
1980- II. Li, Rui, 1981- III. Title.
  TK5103.4.Z48 2011
  621.381'3224–dc23

                                                              2011034132

10 9 8 7 6 5 4 3 2 1

CONTENTS

PREFACE

Microwave filters have been widely used and considered as indispensable building blocks in modern wireless and telecommunication systems. The history of microwave filters can be tracked back before World War II. The early contribution was primarily made to systematically establish two well-known network-oriented design methodologies with resort to the characteristic parameters of image impedances and attenuation functions. In particular, the latter has been widely applied nowadays as the most efficient approach for the synthesis of a variety of microwave filters. By introducing a closed-form transfer function, such as the Chebyshev function, a set of design formulas can be established to instantly determine all the element values involved in a filter topology with specified frequency responses, such as equal-ripple in-band response. Until the 1980s, this approach had been applied as a unique methodology in the design of many standard microwave filters, such as bandpass filters with narrow passband on homogeneously filled waveguide or coaxial transmission lines. Since then, as this approach is highly desired in radiofrequency (RF) and microwave integrated circuits and modules for wireless communication, extensive research has been carried out to explore a variety of microwave filters on inhomogeneous dielectric substrates, such as single-layer printed circuit board (PCB) and multilayer low-temperature cofired ceramic (LTCC). In this context, the conventional synthesis approach is applied to estimate the initial dimensions of a filter without accounting for frequency dispersion, discontinuity effects, and so on. With the help of full-wave simulators, the final layout of a filter is determined through multiple-round electromagnetic simulations on its overall structure toward gradually obtaining the specified filtering performances. Of all these filters, various bandpass filters on a planar configuration have been extensively studied and developed to achieve highly specified filtering functionalities, such as dual-/triple-band and wideband, as required in advanced wireless communications.

Accompanying the recent interest in the development of ultra-wideband (UWB) wireless systems, the required fractional bandwidth has been doubled or significantly widened as compared with those for the traditional narrowband systems. Under the nondispersive lumped-element assumption, the conventional synthesis methodology is no longer capable of designing a bandpass filter covering a wide or ultra-wide passband with a fractional bandwidth of about 110%. Until now, much research has focused on exploring a variety of (ultra)-wideband filters based on the different design approaches. Among them, the technique using the multi-mode resonator (MMR) has been considered one of the most popular solutions in forming a class of (ultra)-wideband filters. As its distinctive feature, multiple resonant modes of the so-called MMR are simultaneously excited to constitute a specified wide or ultra-wide passband. Based on this simple concept, various UWB bandpass filters with different geometrical layouts have been studied since 2005 by us and many other groups in the world. In addition, a direct synthesis method has been recently developed to provide an efficient approach for designing this class of UWB filters. The primary motivation of this book is to summarize the many excellent works in the research topic of wideband filters, and to give a systematic guideline in efficient and accurate design of microwave bandpass filters for wideband application.

In organizing this book, we have attempted to address all the fundamental theory related to the filter design using a resonator with multiple resonant modes. Starting with the introduction of the basic transmission line theory and network analysis, the traditional narrowband filter design theory and analysis of microwave resonators are conducted. Then, we present and characterize a variety of multi-mode resonators (MMRs) with stepped-impedance or loaded-stub configurations using the matured transmission line theory, and further apply them for the design and development of varied microwave wideband filters on single or composite transmission lines. Later on, a direct synthesis approach is presented to provide a series of useful design charts and tables for the proposed MMR-based bandpass filters. After that, we move to give our brief overview on many other works reported by the peers in our society. We hope that this book would be a good reference to the readers who are carrying out their academic research or development work on microwave wideband filters, circuits and antennas.

Herein, we would like to thank all the authors and coauthors of the literatures, listed in the references, for their contributions to this book. In particular, we would like to thank Professor Wolfgang Menzel at University of Ulm in Germany for his close collaboration in this research. As the first author of this book, Professor Zhu would like to give his sincere gratitude to all his other supervised PhD students, Dr Hang Wang, Dr Jing Gao, Dr Sai Wai Wong, Ms Sha Luo, and Mr Teck Beng Lim for their technical contribution to the exploration of various MMR-based filters. Finally, we would like to give our appreciation to Professor Kai Chang at Texas A & M University, Editor-in-Chief of this book series, for his passional invitation. Also, we are pleased to acknowledge the willing and professional cooperation of the publishers. Without their encouragement and assistance, we could not have accomplished this book.

LEI ZHU
SHENG SUN
RUI LI

August 19, 2010

CHAPTER 1

INTRODUCTION

"Wideband" is referred to as a wide operating range of frequencies in microwave engineering, and its relevant technique was initially developed and applied for military communication in the past few decades. In recent years since 2000, unlicensed usage of ultra-wideband (UWB) spectrum has been progressively released globally for short-range wireless communications. It stimulates much interest in exploration of various wideband or UWB techniques for civil applications. As compared with traditional narrow-band communication, wideband or UWB communication has a doubled or extremely wide operating bandwidth so as to bring out its unique feature in enabling high-speed data transfer for short-range wireless connections as well as applications in low data rate, radar, and imaging systems.

Tracking the history of UWB, wireless communication via electromagnetic wave began with transmission and reception of a time-domain pulse signal in an ultra- or very-wide frequency range more than 100 years ago. In 1886, Heinrich Hertz proofed the Maxwell equations via experimental realization of a spark gap transmission, and in 1895, Guglielmo Marconi built up the first radio commutation system in his laboratory in Italy. As a key building block in this wireless system, the antenna was invented and constructed by inducing the radiation through

Microwave Bandpass Filters for Wideband Communications, First Edition. Lei Zhu, Sheng Sun, Rui Li.
© 2012 John Wiley & Sons, Inc. Published 2012 by John Wiley & Sons, Inc.

a spark from a metal plate. One year later, with the joint effort of Marconi and William Preece who was the chief telegraph engineer in Britain, the first worldwide UWB communication system was built up in London in 1896. Hence, two post offices were wirelessly linked through a distance greater than 1 mile.

In the past few decades, significant research progress and achievement on radio frequency (RF) and microwave devices made it possible to develop commercial UWB systems. Meanwhile, extensive research and development activities on UWB technology have been regaining much attention in both academic and industrial aspects due to the recently increased requirement in high-speed and high-data rate communication. As a key component, microwave bandpass filter plays an indispensable role in regulating the limited UWB masks and dominating the frequency functionality of the whole system. In this chapter, a short introduction on the background of UWB technology will be firstly made. After that, the UWB radiation masks and the bandwidth requirements for the UWB bandpass filters will be discussed, followed by a brief review of the recently developed UWB bandpass filters.

1.1   BACKGROUND ON UWB TECHNOLOGY

UWB wireless communication is not a new term, and it was studied in the past for different purposes. Because of low spectral efficiency of the UWB signal generated by a spark-gap transmitter, the narrowband communication systems were much popular instead of UWB systems since 1910s. The fading of UWB research had been continued until the late of 1960s. In 1960, a UWB impulse was rebuilt in the experiment done by Henning Harmuth from the Catholic University of America, and Gerald Ross and K.W. Robins from Sperry Rand Corporation [1]. From 1960s to the 1990s, the UWB technology was restricted to military and defense applications under various classified programs, such as highly secure communications [2]. At that moment, the spectral efficiency was not a major issue, and the spatial resolution was the most important factor in order to improve the accuracy of radar tracking. In 1998, the U.S. Federal Communications Commission (FCC) recognized the significance of UWB technology and released the initial report in February 2002, in which the UWB technology was authorized for the unlicensed commercial uses with different civil applications [3]. Recently, the development of high-speed microprocessors and fast-switching techniques make this UWB technology possible for commercial short-range communications [4–8].

The UWB technology offers a promising solution to the RF spectrum shortage by allowing new services to coexist with current radio systems with minimal or no interference. It transmits extremely short pulses with relatively low energy and occupies an ultra-wide frequency bandwidth. Since frequency is inversely related to time, the short-duration UWB pulses spread their energy across a wide range of frequencies with very low-power spectral density. As a result, it allows the UWB signal to coexist with current radio services, and interferences might not occur. Besides the ability of sharing the frequency spectrum, UWB signals have additional advantages such as,

- Large channel capacity makes UWB system a perfect candidate for short-range, high-data rate wireless applications.
- Low power density that is usually below environment noise ensures communication security.
- The short-duration pulse lowers the sensitivity to multi-path effect.
- Carrierless UWB transmission simplifies the transceiver architecture and thus reduces the cost in design circle and implementation.

1.2 UWB REGULATIONS

To cater for the potential application on market, a few national or international organizations, such as the FCC in the United States, the Electronic Communications Committee (ECC) in Europe, Asia-Pacific Telecommunity (APT), as well as governmental institutes in Japan, Korea, Singapore, and Australia, have realized their own UWB masks in a specific region or country. They have granted a waiver that will lift certain limits on the UWB devices. As mentioned before, the UWB service occupies a very wide bandwidth and shares the frequency spectrum with many other existing services. This means that the existing narrowband radio regulations have to accommodate the UWB rules. Therefore, a framework needs to be generated such that the UWB systems can peacefully coexist with other legacy wireless systems.

1.2.1 FCC Radiation Masks

In fact, the FCC in the United States authorized the unlicensed commercial deployment of UWB technology in February 2002 under a strict power control in 7500 MHz spectrum [5]. The FCC has assigned conservative emission masks between 3.1 and 10.6 GHz for commercial UWB devices—that is, −41.3 dBm/MHz, or 75 nW/MHz—places them at the same level as unintentional radiators such as televisions and computer monitors. Based on the FCC regulations, UWB devices are classified into three major categories: communications, imaging, and vehicular radar. Throughout the full range of UWB applications, applying UWB into short-range high-speed wireless communication is currently considered as one of the most promising areas [7].

For a communication device, the FCC has further assigned different emission limits for indoor and outdoor UWB devices. The spectrum mask for the outdoor devices is 10 dB lower than that for the indoor devices between 1.61 and 3.1 GHz and above 10.6 GHz. The emission limits for the indoor and outdoor UWB devices are shown in Figure 1.1a,b, respectively, where the masks of effective isotropic radiation power (EIRP) are also tabulated in Table 1.1. Herein, we notice that a very wide frequency range has been assigned from 3.1 to 10.6 GHz, indicating a huge wide passband with the fractional bandwidth of 109.5% at the center frequency of 6.85 GHz. In order to reduce the interferences with other narrowband communications, such as global positioning system (GPS), the radiation masks for both of indoor and outdoor systems must be 30 dB lower than those in the UWB band from 0.96 to 1.61 GHz.

According to the FCC regulations, indoor UWB devices must consist of handheld equipments, and their activities should be restricted to peer-to-peer operations inside buildings. The FCC also dictates that no fixed infrastructure can be used for UWB communications in outdoor environments. Therefore, outdoor UWB communications are restricted to handheld devices that can send information only to their associated receivers.

1.2.2 ECC Radiation Masks

In Europe, the European Conference of Postal and Telecommunications Administrations (CEPT) allowed the use of above FCC-specified spectrum for equipments using UWB technology, and the ECC has already made a set of decisions for the devices using UWB technology. As the first mandate issued by the European Commission to CEPT on March 11, 2004, the harmonized use of radio spectrum for UWB applications in the European Union was developed. This ECC decision released the unlicensed using of generic UWB devices below 10.6 GHz. During the ECC meeting on March 24, 2006, the new regulation (ECC/

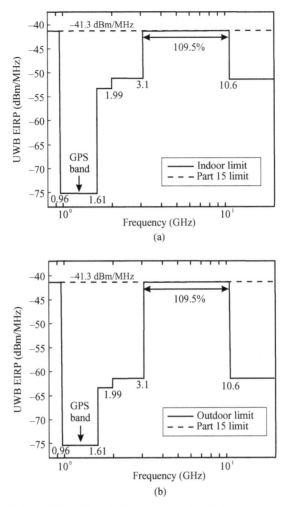

Figure 1.1. FCC-defined UWB radiation masks. (a) Indoor systems. (b) Outdoor systems.

TABLE 1.1 FCC Emission Masks for Indoor and Outdoor UWB Devices

Frequency Range (MHz)	Indoor EIRP (dBm/MHz)	Outdoor EIRP (dBm/MHz)
960–1,610	−75.3	−75.3
1,610–1,990	−53.3	−63.3
1,990–3,100	−51.3	−61.3
3,100–10,600	−41.3	−41.3
Above 10,600	−51.3	−51.3

DEC/(06)04) was introduced in the European Union for devices using UWB technology [9]. The new regulation requires that the UWB applications must also take into account the need of protection for the existing wireless services. That means that UWB devices should provide enough predesigned protection and thus avoid interferences.

The technical requirements specified by ECC in Europe is shown in Figure 1.2 and tabulated in Table 1.2. In addition, the 4.2–4.8 GHz

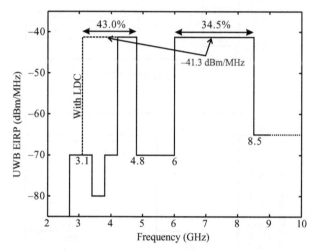

Figure 1.2. ECC-defined UWB radiation masks.

TABLE 1.2   ECC Emission Masks for the Devices Using UWB Technology in Europe

Frequency Range (MHz)	EIRP (dBm/MHz)
below 1600	−90
1600–2700	−85
2700–3400	−70
3400–3800	−80
3800–4200	−70
4200–4800[a]	−70
4800–6000	−70
6000–8500	−41.3
8500–10,600[b]	−65
Above 10,600	−85

[a] Before December 31, 2010, the UWB devices are permitted to operate in the band of 4.2–4.8 GHz with a maximum EIRP spectral density of −41.3 dBm/MHz. The band of 3.1–4.8 GHz has been assigned for the LDC UWB devices.

[b] The covering of the band of 8.5–9 GHz is still under consideration.

frequency band with a maximum mean EIRP spectral density of −41.3 dBm/MHz is for the first generation UWB devices, which is permitted to operate until a fixed cutoff date of December 31, 2010.

As a long-term UWB operation in Europe, the frequency band 6.0–8.5 GHz has been identified without the requirement for additional mitigation. In the decision on December 1, 2006 (ECC/DEC/(06)12), the low band of UWB 3.1–4.8 GHz is also assigned for the UWB devices with low duty cycle (LDC) or detect and avoid (DAA) mitigation techniques. For the frequency band 8.5–9 GHz, ECC has also investigated LDC and DAA mitigation techniques in order to ensure the protection of broadband wireless access (BWA) terminals and applications in the radiolocation services.

Comparing the ECC limits with the FCC limits in Figures 1.1 and 1.2, we can figure out that the UWB emission in Europe is more restrictive than that in United States. Two wide frequency bands with 43.0 and 34.5% have been assigned separately below 10.6 GHz, thus providing more protection to the existing service located besides them.

1.2.3 UWB Definition and Bandwidth

According to the FCC's proposal in 2002 [3], the fractional bandwidth (*FBW*) of a UWB device should be larger than 0.20 or 20%, and the minimum bandwidth (*BW*) limit of 500 MHz at any center frequency should be accommodated, that is,

$$FBW > 20\% \tag{1.1a}$$

$$BW > 500\,\text{MHz}. \tag{1.1b}$$

The formula of *FBW* is defined by using −10 dB emission points and given by

$$FBW = \frac{BW}{f_C} = \frac{2(f_H - f_L)}{f_H + f_L}, \tag{1.2}$$

where f_L and f_H are the lower and upper frequencies of the −10 dB emission points, respectively. The center frequency (f_C) of the UWB transmission was defined as the average of the upper and lower −10 dB points, that is,

$$f_C = \frac{f_H + f_L}{2}. \tag{1.3}$$

TABLE 1.3   UWB Definitions of f_L, f_H, f_C, FBW, and Emission Level for Indoor Devices from Different Areas

Regulations	Frequency Band (GHz) f_L	f_H	Center Frequency (GHz) f_C	Fractional Bandwidth FBW	Emission Level (dBM/MHz)
FCC	3.1	10.6	6.85	109.5%	−41.3
ECC	3.1	4.8	3.95	43.0%	−41.3
	6.0	8.5	7.25	34.5%	−41.3
Japan	3.4	4.8	4.1	34.1%	−41.3
	7.25	10.25	8.75	34.3%	−41.3
Korea	1.0	10.0	5.5	163.6%	−66.5
Singapore	2.2	10.6	6.4	131.25%	−35.0

As an example, consider the UWB radiation masks in Figure 1.1; the f_L and f_H of the defined UWB passband are 3.1 and 10.6 GHz, respectively. Thus, the center frequency and the fractional bandwidth of the UWB passband are calculated as

$$f_C = \frac{f_H + f_L}{2} = 6.85 \, \text{GHz}, \tag{1.4}$$

and

$$FBW = \frac{2(f_H - f_L)}{f_H + f_L} = 109.5\%. \tag{1.5}$$

As mentioned before, the regulations in different regions are different, and the defined UWB frequency bands may be a single band or multiple bands. Table 1.3 concludes some specified UWB definitions for indoor devices in the different regions or countries.

1.3   UWB BANDPASS FILTERS

UWB microwave bandpass filters have recently been receiving enormous attention in both academia and industry for applications in wireless transmission systems. Since the FCC in the United States approved the unlicensed use of the UWB spectrum, researchers in microwave society have been shifting their attention to the develop-

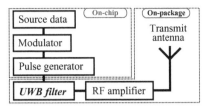

Figure 1.3. A general block diagram of a UWB transmitter system.

ment of UWB bandpass filters challenging the strictly specified FCC emission mask. A general block diagram of a UWB transmitter is depicted in Figure 1.3, where the source data is encoded, modulated, and multiplexed at the chip level, and then the multiplexed pulse is transmitted by a UWB antenna after reshaping and amplifying at the package level.

Similar to many traditional narrowband transmission systems, filter blocks are always needed to remove the unwanted signals and noise from UWB transmission systems. However, unlike the narrowband system discussed in Reference 10, UWB systems spread the desired signals across a very wide frequency range and broadcast them in separate bands. For indoor and hand-held UWB systems, devices must operate with a 10-dB bandwidth in the frequency range of 3.1–10.6 GHz as shown in Figure 1.1. These specifications create a tremendous challenge for a filter designer to design a *FBW* of about 109.5% at the center frequency of 6.85 GHz. As well known, the existing filter theory was established under the assumption of a narrow passband, and it has been found very powerful in the design of microwave filters with various narrow-band filtering responses [11–13]. A wideband filter can be reasonably constructed by cascading a few transmission line resonators through enhanced coupling. Nevertheless, it has still been theoretically difficult to design filters with an ultra-wide passband bandwidth.

Several methodologies have reported to build up a few classes of UWB bandpass filters. The conventional highpass prototype with short-circuited shunt stubs [14] has been commonly adopted in designing bandpass filters with ultra-wide bandwidth [15–20]. Following that, an ultra-wideband bandpass filter was initially presented in Reference 15 and then was systematically optimized by cascading such highpass sections with an active isolation and a lowpass section [16]. However, the resultant filter occupied a large circuitry area. Hence, extensive work has been carried out toward miniaturizing the overall filter size and improving the electrical performance at the same time. In Reference 17, with the electromagnetic bandgap structure formed on the connecting line between two adjacent stubs, the upper stopband was effectively extended and the total circuitry size was miniaturized. Later on, a cross-coupling route was purposely introduced between the input and output ports in References 18 and 19, resulting in enhanced filter selectivity and improving the group delay performance. By adding two quarter-wavelength transmission line sections at two ports and replacing the connecting line by a lowpass structure, a five-pole UWB filter with a widened upper stopband was constructed in Reference 20. This prototype has been recently employed to design a UWB filter on a multilayer liquid-crystal polymer substrate [21–23]. However, in constructing these filters on microstrip line topology, via holes are always required to realize the short-ended terminals for all shunt stubs.

Another major class of UWB bandpass filters based on the so-called multiple-mode resonator (MMR) theory becomes quite popular due to its simple structure and easy design procedure [24–32]. The stepped version of the MMR is also commonly referred to as stepped impedance resonator (SIR). In this book, the term MMR is used to indicate a nonuniform resonator with straight, bent, stepped impedance, or stub-loaded shape. Different from the SIR in narrow-band filter design, the proposed MMR is primarily targeted to excite the multiple resonant modes simultaneously such that an ultra-wide passband can be constructed. The concept of this MMR resonator was firstly introduced in Reference 27, and then implemented on various transmission-line structures [28–37], such as microstrip line (MSL), coplanar waveguide (CPW), hybrid MSL/CPW, MSL/slotline, and so on. One of the main drawbacks of the initial MMR-based bandpass filters is that it has a relatively narrow upper stopband due to the higher-order harmonic resonances. Therefore, a periodically varied electromagnetic bandgap (EBG) structures embedded resonator was proposed in References 35–37 to effectively suppress the harmonic bands while reducing the overall filter size. In parallel, various lowpass or bandstop structures were formed on the MMR so as to provide an alternative way in improving the out-of-band performance as reported in References 32–44. Although the UWB communication with low-power density does not tremendously affect the existing radio systems, the exclusive use of the frequency spectrum, such as wireless local area network (WLAN), may interfere with the UWB communication in certain frequencies. To circumvent this problematic issue, various UWB filters with a single [45–50] or multiple [51] narrow notch bands embedded in the ultra-wide passband were recently developed.

There were also other types of UWB bandpass filters designed with different approaches [52–55]. The resonant modes of a SIR and the split modes of cascading resonators were studied in detail in References 56–57. The resonant peaks of the multistage structure were matched with the nominal Chebyshev poles. The traditional edge-coupled line structure in this type of microstrip line filter may not be strong enough, and it is limited by the tolerance in fabrication process. Therefore, the broadside coupling structures become one solution for this problem [58, 59].

1.4 ORGANIZATION OF THE BOOK

In this first chapter, the background on the UWB technology is briefly discussed. Various UWB regulations are introduced, and the specified fractional bandwidths are also summarized. UWB can be considered as a form of an oldest communication technology. Today, this technology has been released freely for the commercial use in short-range wireless communication. Since UWB communication occupied a very wide frequency band, the necessity in developing a class of wideband filters and limitation in fractional bandwidth of classical bandpass filters are discussed. The research on recent development of various UWB bandpass filters is reviewed at the end of this chapter.

In Chapter 2, an introduction on transmission line theory is given. Based on the telegraph equations, the wave voltages and currents along a transmission line are defined and used to characterize the wave distribution over a length of transmission line. Applying the circuit or network theorems on lumped or distributed elements makes the microwave filter design simpler than those directly using the field theory from the Maxwell's equations. Thus, it is the primary purpose of Chapter 2 to outline the circuit analysis concepts and the two-port network parameters for the design of microwave filters will be needed in the later chapters of this book.

In Chapter 3, the well-known synthesis design of the conventional narrowband bandpass filters is overviewed based on the insertion loss method. The overall design procedure starts with the formation and determination of a lowpass filter prototype via the closed-form transfer functions, and then converts this filter prototype with known network parameters to a microwave filter with varied frequency responses by using frequency transformations. For example, a parallel-coupled line filter is designed from its corresponding lowpass filter prototype and then its lumped-element bandpass network with frequency-independent network parameters. Since all the design formulas are formed in an extremely narrow band around the center frequency, the designed bandpass filter is definitely restricted to be a narrow or moderate passband. Thus, the above limitation motivates us to establish a new method in design of a bandpass filter with a large fractional bandwidth.

In the design procedure of microwave filters, it is important to understand the working principles and analysis methodologies of various microwave resonators since the resonator is a fundamental element in a whole filter block. For this purpose, Chapter 4 focuses itself on the analysis of basic resonators with uniform and nonuniform impedance transmission lines. In particular, multiple resonant modes of a line resonator are extensively studied so as to determine the locations of nearby multiple resonances and to utilize them together in forming a wide passband as expected. The resonant circuits and derivations of Chapter 4 are later applied to the design and implementation of a class of MMR-based UWB bandpass filters.

In Chapter 5, a class of MMR-based wideband bandpass filters with varied configurations and implementation schemes are presented. These filters are formed based on various transmission line structures, inclusive of microstrip-line (MSL), aperture-backed MSL, coplanar waveguide (CPW), hybrid MSL/CPW and MSL/slotline, EBG/LPF-embedded configurations, and so on. After an initial MMR-based UWB parallel-coupled line bandpass filter is proposed and designed, a series of modified designs are carried out with improved filtering performance. In addition, design concerns on the notch-band UWB bandpass filters are discussed in the end.

Synthesis approach is always a preferred and effective way for the design of the microwave filters so as to determine the filter dimensions in a straightforward manner once the filter specifications are given. Therefore, the focus of Chapter 6 is to study the direct synthesis design procedure for MMR-based UWB filters with Chebyshev equal ripple responses in the core UWB passband. After the initial overall dimensions of a UWB bandpass filter are determined, the final filter layout, including complicated frequency dispersions and discontinuity effects, are decided via full-wave electromagnetic simulations.

Besides the MMR-based UWB bandpass filters with different geometric variations, many other types of wideband filters are reviewed in Chapter 7. They are roughly classified into a few categories, such as UWB filters with highpass and lowpass filters, UWB filters with quasi-lumped elements, UWB filters using multilayer structure, and so on. For each category, one or two designs are illustrated in detail for their working mechanisms and design methodologies. Therefore, the readers can have a whole view on current development of UWB filters.

REFERENCES

1 R. Fontana, "A brief history of UWB communications," Online article, http://www.jacksons.net/tac/First%20Term/A_Brief_History_of_UWB_Communications.pdf

2 F. Nekoogar, *Ultra-Wideband Communications: Fundamentals and Applications*, Prentice Hall, Upper Saddle River, NJ, 2005.

3 "Revision of part 15 of the commission's rules regarding ultra-wideband transmission systems," First Note and Order, Federal Communications Commission, ET-Docket 98-153, February 14, 2002.

4 G. R. Aiello and G. D. Rogerson, "Ultra-wideband wireless systems," *IEEE Microw. Mag.* 4(2) (2003) 36–47.

5 L. Q. Yang and G. B. Giannakis, "Ultra-wideband communications: An idea whose time has come," *IEEE Signal Process. Mag.* 21(11) (2004) 26–54.

6 D. Porcino and W. Hirt, "Ultra-wideband radio technology: Potential and challenges ahead," *IEEE Commun. Mag.* 41 (2003) 66–74.

7 S. Roy, J. R. Foerster, V. S. Somayazulu, and D. G. Leeper, "Ultra-wideband radio design: The promise of high-speed, short-range wireless connectivity," *Proc. IEEE* 92(2) (2004) 295–311.

8 R. J. Fontana, "Recent system applications of short-pulse ultra-wideband (UWB) technology," *IEEE Trans. Microwave Theory Tech.* 52(9) (2004) 2087–2104.

9 "ECC Decision of 24 March 2006 amended 6 July 2007 at Constanta on the harmonised conditions for devices using Ultra-Wideband (UWB) technology in bands below 10.6 GHz," Electronic Communications Committee, ECC/DEC/(06)04, amended July 6, 2007.

10 J. D. G. Swanson, "Narrow-band microwave filter design," *IEEE Microw. Mag.* 8(5) (2007) 105–114.

11 G. L. Matthaei, L. Young, and E. M. T. Jones, *Microwave filters, impedance-matching networks, and coupling structures*, Artech House, Dedham, MA, 1980.

12 R. Levy and S. B. Cohn, "A history of microwave filter research, design, and development," *IEEE Trans. Microw. Theory Tech.* 32(9) (1984) 1055–1067.

13 J.-S. Hong and M. J. Lancaster, *Microstrip filters for RF/microwave applications*, Wiley, New York, 2001.

14 R. Levy and L. F. Lind, "Synthesis of symmetrical branch-guide directional couplers," *IEEE Trans. Microw. Theory and Tech.* MTT-16 (1968) 80–89.

15 W.-T. Wong, Y.-S. Lin, C.-H. Wang, and C. H. Chen, "Highly selective microstrip bandpass filters for ultra-wideband (UWB) applications," *Proc. Asia–Pacific Microw. Conf.*, vol. 5, 2005, pp. 2850–2853.

16 R. Gomez-Garcia and J. I. Alonso, "Systematic method for the exact synthesis of ultra-wideband filtering responses using high-pass and low-pass sections," *IEEE Trans. Microw. Theory Tech.* 54(10) (2006) 3751–3764.

17 J. Garcia-Garcia, J. Bonache, and F. Martin, "Application of electromagnetic bandgaps to the design of ultra-wide bandpass filters with good out-of-band performance," *IEEE Trans. Microw. Theory Tech.* 54(12) (2006) 4136–4140.

18 H. Shaman and J.-S. Hong, "A novel ultra-wideband (UWB) bandpass filter (BPF) with pairs of transmission zeroes," *IEEE Microw. Wireless Compon. Lett.* 17(2) (2007) 121–123.

19 H. Shaman and J.-S. Hong, "Input and output cross-coupled wideband bandpass filter," *IEEE Trans. Microw. Theory Tech.* 55(12) (2007) 2562–2568.

20 M. Uhm, K. Kim, and D. S. Filipovic, "Ultra-wideband bandpass filters using quarter-wave short-circuited shunt stubs and quarter-wave series transformers," *IEEE Microw. Wireless Compon. Lett.* 18(10) (2008) 668–670.

21 Z.-C. Hao and J.-S. Hong, "Ultra-wideband bandpass filter using multilayer liquid-crystal-polymer technology," *IEEE Trans. Microw. Theory Tech.* 56(9) (2008) 2095–2100.

22 Z.-C. Hao and J.-S. Hong, "Ultra wideband bandpass filter using embedded stepped impedance resonators on multilayer liquid crystal polymer substrate," *IEEE Microw. Wireless Compon. Lett.* 18(9) (2008) 581–583.

23 Z.-C. Hao and J.-S. Hong, "Compact ultra-wideband bandpass filter using broadside coupled hairpin structures on multilayer liquid crystal polymer substrate," *Electron. Lett.* 44(20) (2008) 1197–1198.

24 W. Menzel, L. Zhu, K. Wu, and F. Bogelsack, "On the design of novel compact broad-band planar filters," *IEEE Trans. Microw. Theory Tech.* 51(2) (2003) 364–370.

25 L. Zhu, H. Bu, and K. Wu, "Broadband and compact multi-mode microstrip bandpass filters using ground plane aperture technique," *IEE Proc. Microw. Antennas Propag.* 149(1) (2002) 71–77.

26 L. Zhu, H. Bu, K. Wu, and M. S. Leong, "Miniaturized multi-pole broadband microstrip bandpass filter: Concept and verification," *30th European Microwave Conf. Proc.*, vol.3, October 2000, Paris, pp. 334–337.

27　L. Zhu, S. Sun, and W. Menzel, "Ultra-wideband (UWB) bandpass filters using multiple-mode resonator," *IEEE Microw. Wireless Compon. Lett.* 15(11) (2005) 796–798.

28　H. Wang, L. Zhu, and W. Menzel, "Ultra-wideband bandpass filter with hybrid microstrip/CPW structure," *IEEE Microw. Wireless Compon. Lett.* 15(12) (2005) 844–846.

29　J. Gao, L. Zhu, W. Menzel, and F. Bogelsack, "Short-circuited CPW multiple-mode resonator for ultra-wideband (UWB) bandpass filter," *IEEE Microw. Wireless Compon. Lett.* 16(3) (2006) 104–106.

30　J. Gao and L. Zhu, "Asymmetric parallel-coupled CPW stages for harmonic suppressed l/4 bandpass filters," *Electron. Lett.* 40(18) (2004) 1122–1123.

31　R. Li and L. Zhu, "Compact UWB bandpass filter using stub-loaded multiple-mode resonator," *IEEE Microw. Wireless Compon. Lett.* 17(1) (2007) 40–42.

32　S. Sun and L. Zhu, "Capacitive-ended interdigital coupled lines for UWB bandpass filters with improved out-of-band performance," *IEEE Microw. Wireless Compon. Lett.* 16(8) (2006) 440–442.

33　R. Li and L. Zhu, "Ultra-wideband (UWB) bandpass filters with hybrid microstrip/slotline structures," *IEEE Microw. Wireless Compon. Lett.* 17(11) (2007) 778–780.

34　R. Li and L. Zhu, "Ultra-wideband microstrip-slotline bandpass filter with enhanced rejection skirts and widened upper stopband," *Electron. Lett.* 43(24) (2007) 1368–1369.

35　S. W. Wong and L. Zhu, "EBG-embedded multiple-mode resonator for UWB bandpass filter with improved upper-stopband performance," *IEEE Microw. Wireless Compon. Lett.* 17(6) (2007) 421–423.

36　S. W. Wong and L. Zhu, "Ultra-wideband bandpass filters with improved out-of-band behavior via embedded electromagnetic-bandgap multimode resonators," Accepted by IET Microwaves," *Antennas and Propagation* 2(8) (2008) 854–862.

37　J.-W. Baik, S.-M. Han, C. Jeong, J. Jeong, and Y.-S. Kim, "Compact ultra-wideband bandpass filter with EBG structure," *IEEE Microw. Wireless Compon. Lett.* 18(10) (2008) 671–673.

38　T. B. Lim, S. Sun, and L. Zhu, "Compact ultra-wideband bandpass filter using harmonic-suppressed multiple-mode resonator," *Electron. Lett.* 43(22) (2007) 1205–1206.

39　C.-W. Tang and M.-G. Chen, "A microstrip ultra-wideband bandpass filter with cascaded broadband bandpass and bandstop filters," *IEEE Trans. Microw. Theory Tech.* 55(11) (2007) 2412–2418.

40　A. Balalem, W. Menzel, J. Machac, and A. Omar, "A simple ultra-wideband suspended stripline bandpass filter with very wide stop-band," *IEEE Microw. Wireless Compon. Lett.* 18(3) (2008) 170–172.

41 N. Thomson and J.-S. Hong, "Compact ultra-wideband microstrip/coplanar waveguide bandpass filter," *IEEE Microw. Wireless Compon. Lett.* 17(3) (2007) 184–186.

42 K. Li, D. Kurita, and T. Matsui, "An ultra-wideband bandpass filter using broadside-coupled microstrip-coplanar waveguide structure," *IEEE MTT-S Int. Dig.*, June 2005, pp. 675–678.

43 K. Li and J.-S. Hong, "Modeling of an ultra-wideband bandpass filtering structure," *Proceeding of APMC2006*, vol. 1, 2006, pp. 37–40.

44 K. Li, Y. Yamamoto, T. Matsui, and O. Hashimoto, "An ultra-wideband (UWB) bandpass filter using broadside-coupled structure and shunt stub with chip capacitor," *Proceeding of APMC2006*, vol. 1, 2006, pp. 41–44.

45 H. Shaman and J.-S. Hong, "Ultra-Wideband (UWB) bandpass filter with embedded band notch structures," *IEEE Microw. Wireless Compon. Lett.* 17(3) (2007) 193–195.

46 H. Shaman and J.-S. Hong, "Asymmetric parallel-coupled lines for notch implementation in UWB filters," *IEEE Microw. Wireless Compon. Lett.* 17(7) (2007) 516–518.

47 K. Li, D. Kurita, and T. Matsui, "Dual-band ultra-wideband bandpass filter," *IEEE MTT-S Int. Dig.*, June 2007, pp. 1193–1196.

48 S. W. Wong and L. Zhu, "Implementation of compact UWB bandpass filter with a notch-band," *IEEE Microw. Wireless Compon. Lett.* 18(1) (2008) 10–12.

49 W. Menzel and P. Feil, "Ultra-wideband (UWB) filter with WLAN notch," *36th European Microw. Conf.*, September 2006, 595–598.

50 G.-M. Yang, R. Jin, C. Vittoria, V. G. Harris, and N. X. Sun, "Small ultra-wideband (UWB) bandpass filter with notched band," *IEEE Microw. Wireless Compon. Lett.* 18(3) (2008) 176–178.

51 K. Li, D. Kurita, and T. Matsui, "UWB bandpass filters with multi notched bands," *36th European Microw. Conf.*, September 2006, 591–594.

52 H. Ishida and K. Araki, "Design and analysis of UWB bandpass filter with ring filter," *IEEE MTT-S Int. Dig.*, vol. 3, June 2004, pp. 1307–1310.

53 C.-L. Hsu, F.-C. Hsu, and J.-T. Kuo, "Microstrip bandpass filters for ultra-wideband (UWB) wireless communications," *IEEE MTT-S Int. Dig.*, June 2005, pp. 679–682.

54 W.-T. Wong, Y.-S. Lin, C.-H. Wang, and C. H. Chen, "Highly selective microstrip bandpass filters for ultra-wideband (UWB) applications," *Proc. Asia–Pacific Microw. Conf.*, vol. 5, December 2005, pp. 2850–2853.

55 W. Menzel, M. S. R. Tito, and L. Zhu, "Low-loss ultra-wideband (UWB) filters using suspended stripline," *Proc. 2005 Asia-Pacific Microw. Conf.*, vol. 4, December 2005, pp. 2148–2151.

56 Y.-C. Chiou, J.-T. Kuo, and E. Cheng, "Broadband quasi-Chebyshev bandpass filter with multimode stepped-impedance resonators (SIRs)," *IEEE Trans. Microw. Theory Tech.* 54 (2006) 3352–3358.

57 J.-T. Kuo, Y.-C. Chiou, and E. Cheng, "High selectivity ultra-wideband (UWB) multimode stepped-impedance resonators (SIRs) bandpass filter with two-layer broadside-coupled structure," *Proc. Asia-Pacific Microw. Conf.*, December 2007, pp. 1–4.

58 A. M. Abbosh, "Planar bandpass filters for ultra-wideband applications," *IEEE Trans. Microw. Theory Tech.* 55(10) (2007) 2262–2269.

59 T.-N. Kuo, S.-C. Lin, and C. H. Chen, "Compact ultra-wideband bandpass filters using composite microstrip-coplanar-waveguide structure," *IEEE Trans. Microw. Theory Tech.* 54(10) (2006) 3772–3778.

CHAPTER 2

TRANSMISSION LINE CONCEPTS AND NETWORKS

2.1 INTRODUCTION

This book addresses the design of microwave wideband bandpass filters. These filters are normally constructed using a variety of transmission lines, for example, microstrip line (MSL), coplanar waveguide (CPW), slotline (SL), and so on. It has been commonly recognized that transmission line model is the first and the most simplified choice for analyzing and synthesizing transmission line filters. By employing the distributed and/or lumped elements, polynomial-based transfer function could be exactly or approximately represented from the network synthesis. However, in practical design, both circuit- and EM-based optimization procedures have to be carried out to take into account the frequency dispersion, losses, discontinuities, and other parasitic effects. Besides the network synthesis, transmission line theory is also necessary in characterizing the resonant behavior of a line resonator, and it transfers a complex field analysis to a simple circuit theory. The most common filter-type structures can be classified as a generalized two-port network. If the input and output parameters are known, the transmission poles and zeros of the filters can be theoretically deter-

Microwave Bandpass Filters for Wideband Communications, First Edition. Lei Zhu, Sheng Sun, Rui Li.
© 2012 John Wiley & Sons, Inc. Published 2012 by John Wiley & Sons, Inc.

mined by performing the network analysis using impedance-, admittance-, scattering- or *ABCD*-matrix parameters. Therefore, the transmission line concept and relevant network parameters are indispensable in analyzing and characterizing a variety of passive microwave components, including filters [1–3]. Thus, the purpose of this chapter is to give a brief introduction on the basic transmission line theory, two-port microwave network, and other relevant theories that would be useful in the design of microwave filters.

2.2 TRANSMISSION LINE THEORY

2.2.1 Basic Transmission Line Model

Unlike circuit theory with much smaller electrical size, transmission line theory assumes that the line length may be a comparable or multiple sections of a wavelength. As a unique property of transmission line, wave propagation along the transmission line can be represented by variable voltages and currents over the line length. Thus, the simple but powerful circuit theory can still be utilized for the analysis if the electrical length is small enough in comparison with wavelength. For a per-unit-length (PUL) transmission line model of a two-wire line, a short increment of line with length Δz can be modeled as a lumped-element equivalent circuit, as shown in Figure 2.1. The PUL series resistance R and PUL shunt conductance G represent two primary transmission losses due to finite conductivity of the conductors and nonzero loss tangent of dielectric substrates, respectively. The PUL series inductance L and shunt capacitance C represent the inductance of the conductors and the capacitance between two conductors, respectively. By employing Kirchhoff's law, the relationship between voltage $V(z)$ and current $I(z)$ in Figure 2.1 can be derived as

$$\frac{dV(z)}{dz} = -(R + j\omega L)I(z) \qquad (2.1a)$$

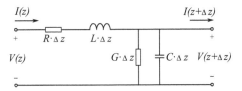

Figure 2.1. Lumped-element equivalent circuit for a transmission line with short length Δz.

$$\frac{dI(z)}{dz} = -(G + j\omega C)V(z). \tag{2.1b}$$

Solving Equation (2.1a,b) gives the following wave equations for $V(z)$ and $I(z)$:

$$\frac{d^2V(z)}{dz^2} - \gamma^2 V(z) = 0, \tag{2.2a}$$

$$\frac{d^2I(z)}{dz^2} - \gamma^2 I(z) = 0, \tag{2.2b}$$

where $\gamma = \alpha + j\beta = \sqrt{(R + j\omega L)(G + j\omega C)}$ is the complex propagation constant, α and β are the attenuation and phase constants, respectively. For an ideal lossless line, $R = G = 0$. The attenuation constant α is zero, and the propagation constant becomes purely imaginary as

$$\gamma = j\beta = j\omega\sqrt{LC}. \tag{2.3}$$

Then, the characteristic impedance, Z_0, can be defined as

$$Z_0 = \sqrt{\frac{L}{C}}. \tag{2.4}$$

If the line has sufficiently small conductor loss and dielectric loss, we can assume that $R \ll \omega L$ and $G \ll \omega C$. Thus, the complex propagation constant can be simplified as

$$\begin{aligned}
\gamma &= \sqrt{(R + j\omega L)(G + j\omega C)} \\
&= j\omega\sqrt{LC}\sqrt{1 - j\left(\frac{R}{\omega L} + \frac{G}{\omega C}\right)} \\
&\approx \frac{R\sqrt{\dfrac{C}{L}} + G\sqrt{\dfrac{L}{C}}}{2} + j\omega\sqrt{LC}.
\end{aligned} \tag{2.5}$$

Strictly speaking, all the transmission lines in practice hold nonzero conductor and/or dielectric losses. However, at microwave or high frequencies, these losses are negligibly small, so the attenuation constant α or the real part of Equation (2.5) can be totally ignored. Thus, the propagation constant γ is approximated as the same as that in the lossless case in Equation (2.3), and the characteristic impedance Z_0 becomes a real number that is the same as Equation (2.4):

$$Z_0 = \sqrt{\frac{(R+j\omega L)}{(G+j\omega C)}} \approx \sqrt{\frac{L}{C}}. \qquad (2.6)$$

Applying Equations (2.3) and (2.4) to the wave equations (Eq. 2.2) gives a set of general traveling wave solutions for the lossless case:

$$V(z) = V_o^+ e^{-j\beta z} + V_o^- e^{j\beta z} \qquad (2.7a)$$

$$I(z) = I_o^+ e^{-j\beta z} + I_o^- e^{j\beta z} = \frac{V_o^+}{Z_0} e^{-j\beta z} - \frac{V_o^-}{Z_0} e^{j\beta z}, \qquad (2.7b)$$

where the $e^{-j\beta z}$ term represents a forward or an incident wave in the $+z$ direction with the voltage amplitude V_o^+, whereas the $e^{j\beta z}$ term represents a backward or a reflected wave in the $-z$ direction with the voltage amplitude V_o^-. That means that the total voltage and current at any position along the line can be written as a linear superposition of the forward and backward waves. The negative sign in Equation (2.7b) indicates opposite direction of the backward- and forward-wave currents at the same position.

As the other two parameters of a transmission line, the guided-wavelength and phase velocity are given as a function of the above two PUL parameters L and C as

$$\lambda = \frac{2\pi}{\beta} = \frac{2\pi}{\omega\sqrt{LC}} \qquad (2.8a)$$

$$v_p = \frac{\omega}{\beta} = \frac{1}{\sqrt{LC}}. \qquad (2.8b)$$

2.2.2 Lossless Terminated Transmission Lines

As a basic property of the travelling wave, a forward wave in a transmission line must be reflected by a mismatched termination, resulting in a backward wave. Figure 2.2 shows the schematic of a lossless

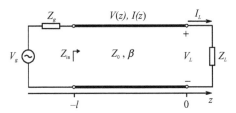

Figure 2.2. A lossless transmission line terminated in an arbitrary load impedance Z_L.

transmission line with a length of l that is terminated in an arbitrary load impedance Z_L.

From Equation (2.7), the load impedance Z_L can be expressed as a ratio of the terminal voltage and current at $z = 0$ as

$$Z_L = \frac{V(0)}{I(0)} = \frac{V_o^+ + V_o^-}{V_o^+ - V_o^-} Z_0. \tag{2.9}$$

The ratio of the backward wave voltage to the forward wave voltage at $z = 0$ is defined as the voltage reflection coefficient, Γ_V, whereas the ratio of the backward wave current to the forward wave current at $z = 0$ is defined as the current reflection coefficient, Γ_I, as follows:

$$\Gamma_V = \frac{V_o^-}{V_o^+} = \frac{Z_L - Z_0}{Z_L + Z_0} \tag{2.10a}$$

$$\Gamma_I = \frac{I_o^-}{I_o^+} = \frac{Y_L - Y_0}{Y_L + Y_0}, \tag{2.10b}$$

where $Y_L = 1/Z_L$ and $Y_0 = 1/Z_0$. From Equation (2.10), it is interesting to note that the current reflection coefficient equals to the negative voltage reflection coefficient at the load. In this book, we will only use the voltage reflection coefficient. In order to avoid unnecessary confusion of these two definitions, the symbol of Γ_V is thus simplified as Γ.

The total voltage and current at the source position, that is, $z = -l$, of the line can be derived as

$$V(-l) = V_o^+ e^{j\beta l} + V_o^- e^{-j\beta l} = V_o^+ e^{j\beta l}[1 + \Gamma e^{-2j\beta l}] \tag{2.11a}$$

$$I(-l) = \frac{V_o^+}{Z_0} e^{j\beta l} - \frac{V_o^-}{Z_0} e^{-j\beta l} = \frac{V_o^+}{Z_0} e^{j\beta l}[1 - \Gamma e^{-2j\beta l}], \tag{2.11b}$$

where Γ is the voltage reflection coefficient at $z = 0$. From Equation (2.11), the input impedance looking toward the load at $z = -l$ is defined and derived as

$$Z_{in} = \frac{V(-l)}{I(-l)} = Z_0 \frac{V_o^+ e^{j\beta l} + V_o^- e^{-j\beta l}}{V_o^+ e^{j\beta l} - V_o^- e^{-j\beta l}} = Z_0 \frac{e^{j\beta l} + \Gamma e^{-j\beta l}}{e^{j\beta l} - \Gamma e^{-j\beta l}} \tag{2.12}$$

Substituting Equation (2.10a) into Equation (2.12), the input impedance can then be written as

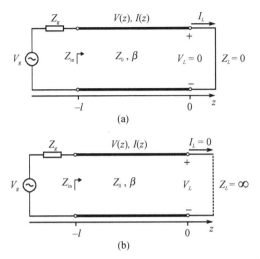

Figure 2.3. A transmission line terminated in short and open circuits. (a) Short-circuited line. (b) Open-circuited line.

$$
\begin{aligned}
Z_{in} &= Z_0 \frac{(Z_L + Z_0)e^{j\beta l} + (Z_L - Z_0)e^{-j\beta l}}{(Z_L + Z_0)e^{j\beta l} - (Z_L - Z_0)e^{-j\beta l}} \\
&= Z_0 \frac{Z_L(e^{j\beta l} + e^{-j\beta l}) + Z_0(e^{j\beta l} - e^{-j\beta l})}{Z_0(e^{j\beta l} + e^{-j\beta l}) + Z_L(e^{j\beta l} - e^{-j\beta l})} \\
&= Z_0 \frac{Z_L \cos \beta l + jZ_0 \sin \beta l}{Z_0 \cos \beta l + jZ_L \sin \beta l} \\
&= Z_0 \frac{Z_L + jZ_0 \tan \beta l}{Z_0 + jZ_L \tan \beta l}.
\end{aligned}
\tag{2.13}
$$

Now, let us consider two special cases that often appear in the analysis procedure of our work herein. Figure 2.3a,b show two special cases that a line is terminated in a short and an open circuit, respectively. For the short-circuited line in Figure 2.3a, the load impedance Z_L becomes zero. So, we can get $\Gamma = -1$ from Equation (2.10a) and the following input impedance from Equation (2.13) as

$$
Z_{in} = jZ_0 \tan \beta l. \tag{2.14}
$$

For the open-circuited line with an infinite load impedance Z_L shown in Figure 2.3b, the reflection coefficient becomes $\Gamma = 1$, and the input impedance is deduced as

$$
Z_{in} = -jZ_0 \cot \beta l. \tag{2.15}
$$

EXAMPLE 2.1   Determine the voltage and current at any point along the transmission line ($z < 0$) as shown in Figure 2.2 and plot it as a function of z. Assuming $V_g = 10\,\text{V}$, $Z_g = Z_0 = 100\,\Omega$, $Z_L = 50 + j25$ Ω, and $l = 1.5\lambda$.

Solution

From Equations (2.8a), (2.10a), and (2.13),

$$\Gamma = \frac{Z_L - Z_0}{Z_L + Z_0} = \frac{-50 + j25}{150 + j25} = -0.297 + j0.216,$$

and

$$Z_{\text{in}} = Z_0 \frac{Z_L + jZ_0 \tan \beta l}{Z_0 + jZ_L \tan \beta l} = Z_0 \frac{Z_L + jZ_0 \tan\left(\dfrac{2\pi}{\lambda} \times \dfrac{3}{2}\lambda\right)}{Z_0 + jZ_L \tan\left(\dfrac{2\pi}{\lambda} \times \dfrac{3}{2}\lambda\right)} = Z_L.$$

Then, the voltage at the input port, V_{in}, can be calculated as

$$V_{\text{in}} = \frac{V_g}{Z_g + Z_{\text{in}}} Z_{\text{in}} = \frac{V_g}{Z_g + Z_L} Z_L = 10\frac{50 + j25}{150 + j25} = 3.513 + j1.081.$$

From Equation (2.11a),

$$V_o^+ = \frac{V_{\text{in}}}{e^{j\beta l}\left(1 + \Gamma e^{-2j\beta l}\right)}.$$

Thus, we have

$$V(z) = V_o^+ e^{-j\beta z} + V_o^- e^{j\beta z} = \frac{V_{\text{in}}\left(e^{-j\beta z} + \Gamma e^{j\beta z}\right)}{e^{j\beta l}\left(1 + \Gamma e^{-2j\beta l}\right)}.$$

The magnitude of the voltage $|V(z)|$ is then plotted in Figure 2.4, and it is demonstrated as a periodical function of the position z with the periodicity of 0.5λ.

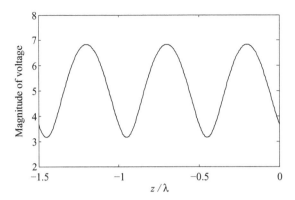

Figure 2.4. The magnitude of the total voltage wave along a 1.5λ line.

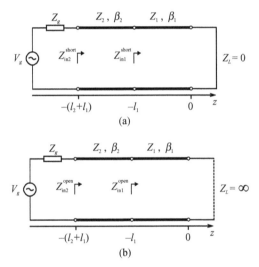

Figure 2.5. Two-section transmission lines terminated in short and open circuits. (a) Short-circuited line. (b) Open-circuited line.

EXAMPLE 2.2 Determine the input impedances of the two-section short- and open-circuited lossless transmission lines as shown in Figure 2.5.

Solution

For a short-circuited line, we can found the input impedance at $z = -l$ from Equation (2.15) as

$$Z_{\text{in1}}^{\text{short}} = jZ_1 \tan \beta_1 l_1.$$

Substituting it into Equation (2.13) gives

$$
\begin{aligned}
Z_{in2}^{short} &= Z_2 \frac{Z_{in1}^{short} + jZ_2 \tan \beta_2 l_2}{Z_2 + jZ_{in1}^{short} \tan \beta_2 l_2} \\
&= Z_2 \frac{jZ_1 \tan \beta_1 l_1 + jZ_2 \tan \beta_2 l_2}{Z_2 + jZ_1 \tan \beta_1 l_1 \times j \tan \beta_2 l_2}. \\
&= jZ_2 \frac{Z_1 \tan \beta_1 l_1 + Z_2 \tan \beta_2 l_2}{Z_2 - Z_1 \tan \beta_1 l_1 \tan \beta_2 l_2}
\end{aligned}
$$

Similarly, for the case of an open-circuited line, we have

$$
Z_{in1}^{open} = -jZ_1 \cot \beta_1 l_1.
$$

Then,

$$
\begin{aligned}
Z_{in2}^{open} &= Z_2 \frac{Z_{in1}^{open} + jZ_2 \tan \beta_2 l_2}{Z_2 + jZ_{in1}^{open} \tan \beta_2 l_2} \\
&= Z_2 \frac{-jZ_1 \cot \beta_1 l_1 + jZ_2 \tan \beta_2 l_2}{Z_2 + j \tan \beta_2 l_2 \times (-jZ_1 \cot \beta_1 l_1)}. \\
&= -jZ_2 \frac{Z_1 - Z_2 \tan \beta_1 l_1 \tan \beta_2 l_2}{Z_2 \tan \beta_1 l_1 + Z_1 \tan \beta_2 l_2}
\end{aligned}
$$

From the results, we note that both two input impedances are still purely reactive as those of the one-section lines. These derivations are useful in our later design, and they will be applied in the analysis of transmission line resonators in Chapter 4.

2.3   MICROWAVE NETWORK PARAMETERS

2.3.1   Scattering Parameters for Two-Port Network

The scattering parameters or S-parameters are very popular for describing a microwave network. In this section, we will briefly introduce the definition of S-parameters for a two-port network. For a multiport network, the detailed discussion can be found in Reference 4. Figure 2.6 shows a two-port network with real characteristic impedance at each port (Z_{01} and Z_{02}). The backward waves at each port are expressed as the response of the forward waves as follows:

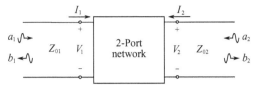

Figure 2.6. A two-port network with the normalized forward and backward waves at two ports.

$$b_1 = a_1 S_{11} + a_2 S_{12} \tag{2.16a}$$

$$b_2 = a_1 S_{21} + a_2 S_{22}. \tag{2.16b}$$

In matrix notation, this set of equations can be represented as

$$[b] = [S][a], \tag{2.17}$$

where a and b are 1×2 matrices of the normalized forward and backward waves as

$$[a] = \begin{bmatrix} a_1 \\ a_2 \end{bmatrix} = \begin{bmatrix} V_1^+ / \sqrt{Z_{01}} \\ V_2^+ / \sqrt{Z_{02}} \end{bmatrix} \tag{2.18a}$$

$$[b] = \begin{bmatrix} b_1 \\ b_2 \end{bmatrix} = \begin{bmatrix} V_1^- / \sqrt{Z_{01}} \\ V_2^- / \sqrt{Z_{02}} \end{bmatrix}, \tag{2.18b}$$

and the scattering matrix is a 2×2 matrix as

$$[S] = \begin{bmatrix} S_{11} & S_{12} \\ S_{21} & S_{22} \end{bmatrix}. \tag{2.19}$$

Then, the scattering parameters, S_{ij} $(i, j = 1, 2)$, are given by

$$S_{11} = \frac{b_1}{a_1}\bigg|_{a_2=0} \quad S_{12} = \frac{b_1}{a_2}\bigg|_{a_1=0}$$

$$S_{21} = \frac{b_2}{a_1}\bigg|_{a_2=0} \quad S_{22} = \frac{b_2}{a_2}\bigg|_{a_1=0}. \tag{2.20}$$

From Equation (2.7), the total voltage and current waves of the n-th port $(n = 1, 2)$ can be derived as

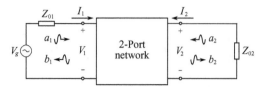

Figure 2.7. A two-port network connected with source and load impedances.

$$V_n^+ = \frac{1}{2}(V_n + Z_{0n}I_n),$$
(2.21a)

$$V_n^- = \frac{1}{2}(V_n - Z_{0n}I_n).$$
(2.21b)

Thus, we have the normalized forward and backward waves as

$$a_n = \frac{V_n^+}{\sqrt{Z_{0n}}} = \frac{1}{2}\left(\frac{V_n}{\sqrt{Z_{0n}}} + \sqrt{Z_{0n}}I_n\right)$$
(2.22a)

$$b_n = \frac{V_n^-}{\sqrt{Z_{0n}}} = \frac{1}{2}\left(\frac{V_n}{\sqrt{Z_{0n}}} - \sqrt{Z_{0n}}I_n\right).$$
(2.22b)

If the two-port network in Figure 2.6 is terminated by the generator V_g with an internal impedance Z_{01} at port 1 and the load impedance Z_{02} at port 2, the total voltage at port 1, shown in Figure 2.7, can be expressed as

$$V_1 = V_g - Z_{01}I_1.$$
(2.23)

Substituting Equation (2.23) into Equation (2.22a) gives

$$a_1 = \frac{V_g}{2\sqrt{Z_{01}}},$$
(2.24)

and the available power generated at port 1 is then given by

$$P_{\text{Avail}} = \frac{1}{2}a_1a_1^* = \frac{|V_g|^2}{8Z_{01}}.$$
(2.25)

Under the matching condition at port 2, $a_2 = 0$ is deduced. From Equation (2.22a,b),

$$0 = \frac{1}{2}\left(\frac{V_2}{\sqrt{Z_{02}}} + \sqrt{Z_{02}}\,I_2\right) \qquad (2.26a)$$

$$b_2 = \frac{1}{2}\left(\frac{V_2}{\sqrt{Z_{02}}} - \sqrt{Z_{02}}\,I_2\right), \qquad (2.26b)$$

or

$$b_2 = \frac{V_2}{\sqrt{Z_{02}}}. \qquad (2.27)$$

Then, we have the power delivered to the load at port 2 as

$$P_L = \frac{1}{2}b_2 b_2^* = \frac{|V_2|^2}{2Z_{02}}. \qquad (2.28)$$

Now the power transfer function of the two-port network of Figure 2.7 is

$$\frac{P_{\text{Avail}}}{P_L} = \frac{\frac{1}{2}a_1 a_1^*}{\frac{1}{2}b_2 b_2^*} = \frac{Z_{02}}{4Z_{01}}\left|\frac{V_g}{V_2}\right|^2. \qquad (2.29)$$

Substituting Equations (2.24) and (2.27) into Equation (2.20c) gives

$$S_{21} = \frac{b_2}{a_1}\bigg|_{a_2=0} = 2\sqrt{\frac{Z_{01}}{Z_{02}}}\left(\frac{V_2}{V_g}\right), \qquad (2.30)$$

and from Equation (2.22) at port 1 and Equation (2.20a), we can find

$$S_{11} = \frac{b_1}{a_1}\bigg|_{a_2=0} = \frac{\dfrac{V_1}{\sqrt{Z_{01}}} - \sqrt{Z_{01}}\,I_1}{\dfrac{V_1}{\sqrt{Z_{01}}} + \sqrt{Z_{01}}\,I_1} = \frac{Z_{in} - Z_{01}}{Z_{in} + Z_{01}}. \qquad (2.31)$$

As usual, S_{21} is called *transmission coefficient* or *voltage gain* of a two-port network, while S_{11} is called *reflection coefficient*. The relationship between the transmission coefficient and the power transfer function can be derived as

$$\frac{P_{\text{Avail}}}{P_L} = \frac{1}{|S_{21}|^2}. \qquad (2.32)$$

In filter design, Equations (2.29) and (2.30) are often expressed in the unit of decibels (dB), leading to the definition of the *insertion loss* (L_A) as

$$L_A = 10\log_{10}\left(\frac{P_{\text{Avail}}}{P_L}\right) \quad (\text{dB})$$

$$= -20\log_{10}|S_{21}| \quad (\text{dB}).$$

(2.33)

For a passive network with varied properties, we can have the following sets of equations:

$$S_{12} = S_{21} \text{ (If network is reciprocal)} \tag{2.34a}$$

$$S_{11} = S_{22} \text{ (If network is symmetrical)} \tag{2.34b}$$

$$\begin{cases} |S_{21}|^2 + |S_{11}|^2 = 1 \\ |S_{12}|^2 + |S_{22}|^2 = 1 \end{cases} \text{(If network is passive and lossless).} \tag{2.34c}$$

2.3.2 [Z] and [Y] Matrices

In Figure 2.7, the relation between the voltages and currents at each port can be defined by an impedance network through

$$V_1 = Z_{11}I_1 + Z_{12}I_2, \tag{2.35a}$$

$$V_2 = Z_{21}I_1 + Z_{22}I_2. \tag{2.35b}$$

where the four impedance parameters can be denoted in a matrix, which is known as impedance matrix or Z-matrix as

$$[Z] = \begin{bmatrix} Z_{11} & Z_{12} \\ Z_{21} & Z_{22} \end{bmatrix}. \tag{2.36}$$

Similar to Equation (2.20), each entry of $[Z]$ can be obtained by setting $I_n = 0$ ($n = 1, 2$), which indicates a perfect open-circuit at port n:

$$Z_{11} = \left.\frac{V_1}{I_1}\right|_{I_2=0} \quad Z_{12} = \left.\frac{V_1}{I_2}\right|_{I_1=0}$$

$$Z_{21} = \left.\frac{V_2}{I_1}\right|_{I_2=0} \quad Z_{22} = \left.\frac{V_2}{I_2}\right|_{I_1=0}$$

(2.37)

Thus, four entries of $[Z]$ are also known as open-circuit impedances, and $[Z]$ is so-called an open-circuit impedance matrix. Due to the property of impedance, $[Z]$ is often employed in the calculation of series network connection. For the parallel network connection, its inversed matrix has to be adopted, which is known as a short-circuit admittance matrix $[Y]$ or Y-matrix

$$[Y]=[Z]^{-1}=\begin{bmatrix} Y_{11} & Y_{12} \\ Y_{21} & Y_{22} \end{bmatrix}. \tag{2.38}$$

Contrary to open-circuit impedance parameters in Equation (2.37), the short-circuit admittance parameters can be obtained by having the short-circuit definition at port n:

$$
\begin{aligned}
Y_{11} &= \frac{I_1}{V_1}\bigg|_{V_2=0} & Y_{12} &= \frac{I_1}{V_2}\bigg|_{V_1=0} \\
Y_{21} &= \frac{I_2}{V_1}\bigg|_{V_2=0} & Y_{22} &= \frac{I_2}{V_2}\bigg|_{V_1=0}.
\end{aligned}
\tag{2.39}
$$

For a reciprocal network, $Z_{12}=Z_{21}$ and $Y_{12}=Y_{21}$ are valid. If the network is also symmetrical, $Z_{11}=Z_{22}$ and $Y_{11}=Y_{22}$ are deduced. For a lossless network, both open-circuit impedance and short-circuit admittance parameters are all purely imaginary. According to Equation (2.38), the conversion between the Z- and Y-matrices can be expanded as

$$[Y]=[Z]^{-1}=\frac{1}{Z_{11}Z_{22}-Z_{12}Z_{21}}\begin{bmatrix} Z_{22} & -Z_{21} \\ -Z_{12} & Z_{11} \end{bmatrix}. \tag{2.40}$$

Furthermore, equivalent T and π networks as shown in Figures 2.8 and 2.9 can be derived to characterize any two-port network in terms of the entries of Z- and Y-matrix. For a two-port T network shown in Figure 2.8a, there are two series impedances (Z_1 and Z_2) and one shunt impedance (Z_3). All these three impedance parameters may be simply

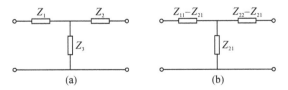

(a)   (b)

Figure 2.8. Equivalent T-network for a two-port junction. (a) Using branch impedances. (b) Using impedance parameters of Z-matrix.

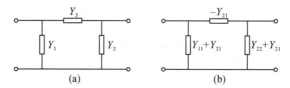

Figure 2.9. Equivalent π-network for a two-port junction. (a) Using branch impedances. (b) Using impedance parameters of Y-matrix.

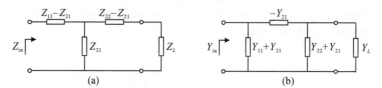

Figure 2.10. Definitions of input parameters for terminated network of a two-port junction. (a) Input impedance. (b) Input admittance.

expressed as a function of three open-circuit impedances of a Z-matrix by using the definition in Equation (2.37)

$$Z_1 = Z_{11} - Z_{21} \qquad (2.41a)$$

$$Z_2 = Z_{22} - Z_{21} \qquad (2.41b)$$

$$Z_3 = Z_{21}, \qquad (2.41c)$$

The π-network depicted in Figure 2.9a consists of two shunt admittances (Y_1 and Y_2) and one series admittance (Y_3). Following the short-circuit definition in Equation (2.39), these three admittances may be expressed in terms of three short-circuit admittances as

$$Y_1 = Y_{11} + Y_{21}, \qquad (2.42a)$$

$$Y_2 = Y_{22} + Y_{21}, \qquad (2.42b)$$

$$Y_3 = -Y_{21}. \qquad (2.42c)$$

Equivalent circuits of the two-port junction in terms of their open-circuit impedances and short-circuit admittances can thus be derived, and they are illustrated in Figures 2.8b and 2.9b, respectively. Input impedance (Z_{in}) and admittance (Y_{in}) of a two-port junction can be then defined by terminating one of the two ports with a load as shown in Figure 2.10. From Figure 2.10a, we can deduce the relations of port voltages and currents as

$$V_1 = Z_{11}I_1 + Z_{12}I_2, \qquad (2.43a)$$

$$-I_2 Z_L = Z_{21}I_1 + Z_{22}I_2. \qquad (2.43b)$$

Solving Equation (2.43) by canceling I_2 gives Z_{in} as

$$Z_{in} = \frac{V_1}{I_1} = Z_{11} - \frac{Z_{12}Z_{21}}{Z_L + Z_{22}}. \qquad (2.44)$$

Similarly, the input admittance of the network shown in Figure 2.10b can be derived as

$$Y_{in} = \frac{I_1}{Y_1} = Y_{11} - \frac{Y_{12}Y_{21}}{Y_L + Y_{22}}. \qquad (2.45)$$

2.3.3 *ABCD* Parameters

Microwave networks can be usually characterized by the S-, Z-, and Y-matrices. For the connection of two-port networks, Z- and Y-matrices can be readily used for the series and parallel combinations. In practice, many filters and other components are actually represented by a series of cascaded two-port networks. Therefore, it is necessary to describe a network by a transmission (T-matrix) or $ABCD$ matrix. Thus, an $ABCD$ matrix of the overall circuit block can be derived by the matrix product of several $ABCD$ matrices of the cascaded two-port networks. For a filter block, the power transfer function in Equation (2.29) or the transmission coefficient in Equation (2.30) can be then obtained based on the chain property of the $ABCD$ matrix.

Let us take a look at a two-port network in Figure 2.11. The relationship between the voltage and current at port 1 (V_1 and I_1) and the voltage and current at port 2 (V_2 and I_2) can be defined via the $ABCD$-matrix as

$$\begin{bmatrix} V_1 \\ I_1 \end{bmatrix} = \begin{bmatrix} A & B \\ C & D \end{bmatrix} \begin{bmatrix} V_2 \\ I_2 \end{bmatrix}, \qquad (2.46)$$

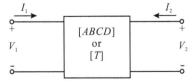

Figure 2.11. A two-port network described by $ABCD$ matrix.

where each entry of the *ABCD*-matrix can be deduced in terms of the *Z*-matrix or *Y*-matrix, which will be discussed in the next section. In order to directly characterize the cascaded connection of two networks, the current flow direction at port 2 has to be reversed as shown in Figures 2.6 and 2.7:

$$\begin{bmatrix} V_1 \\ I_1 \end{bmatrix} = \begin{bmatrix} A_1 & B_1 \\ C_1 & D_1 \end{bmatrix} \begin{bmatrix} V_2 \\ I_2 \end{bmatrix} \tag{2.47a}$$

$$\begin{bmatrix} V_2 \\ I_2 \end{bmatrix} = \begin{bmatrix} A_2 & B_2 \\ C_2 & D_2 \end{bmatrix} \begin{bmatrix} V_3 \\ I_3 \end{bmatrix} \tag{2.47b}$$

$$\begin{bmatrix} V_n \\ I_n \end{bmatrix} = \begin{bmatrix} A_n & B_n \\ C_n & D_n \end{bmatrix} \begin{bmatrix} V_{n+1} \\ I_{n+1} \end{bmatrix} \tag{2.47c}$$

Now, let us look at a general cascaded network as shown in Figure 2.12. Substituting Equation (2.47a) in Equation (2.47b) and repeating this substitution until Equation (2.47c) can lead to

$$\begin{bmatrix} V_1 \\ I_1 \end{bmatrix} = \begin{bmatrix} A_1 & B_1 \\ C_1 & D_1 \end{bmatrix} \begin{bmatrix} A_2 & B_2 \\ C_2 & D_2 \end{bmatrix} \cdots \begin{bmatrix} A_n & B_n \\ C_n & D_n \end{bmatrix} \begin{bmatrix} V_{n+1} \\ I_{n+1} \end{bmatrix}, \tag{2.48}$$

Thus, the port voltage and current at port 1 (left terminal), V_1 and I_1, can be directly derived as a function of those at port $n + 1$ (right terminal), V_{n+1} and I_{n+1}. As two special characteristics, the *ABCD* matrix holds $A = D$ for a symmetrical network and for a reciprocal network, we can derive

$$AD - BC = 1. \tag{2.49}$$

For a lossless network, A and D are purely real, while B and C are both purely imaginary. If the two-port network shown in Figure 2.13 is terminated with a load impedance Z_L at port 2, the input impedance at port 1 can be then derived based on the definition of (2.26) as

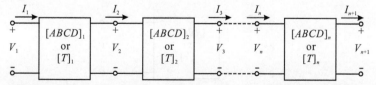

Figure 2.12. A general cascade connection of n two-port networks.

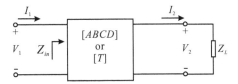

Figure 2.13. A two-port network terminated in a load.

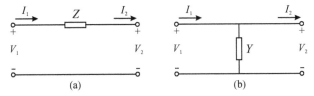

Figure 2.14. Two circuit elements defined by $ABCD$ matrix. (a) Series circuit element. (b) Shunt circuit element.

$$Z_{in} = \frac{V_1}{I_1} = \frac{AV_2 + BI_2}{CV_2 + DI_2}$$
$$= \frac{AV_2/I_2 + B}{CV_2/I_2 + D} \tag{2.50}$$
$$= \frac{AZ_L + B}{CZ_L + D}.$$

Equivalent T and π networks in Figures 2.8 and 2.9 have both series and shunt elements, so it can be figured out that all the cascaded networks can be generally expressed in terms of its overall $ABCD$ matrix, which is derived by multiplying the $ABCD$ matrices of these elements. Based on the Kirchhoff's voltage and current laws, the port voltages and currents for the series and shunt circuit elements in Figure 2.14 can be obtained as

$$\left.\begin{array}{l} V_1 - V_2 = I_2 Z \\ I_1 = I_2 \end{array}\right\} \text{(for series circuit element),} \tag{2.51a}$$

$$\left.\begin{array}{l} I_1 - I_2 = V_2 Y \\ V_1 = V_2 \end{array}\right\} \text{(for shunt circuit element).} \tag{2.51b}$$

Rearranging Equation (2.51a,b) in matrix denotation gives

$$\begin{bmatrix} V_1 \\ I_1 \end{bmatrix} = \begin{bmatrix} 1 & Z \\ 0 & 1 \end{bmatrix} \begin{bmatrix} V_2 \\ I_2 \end{bmatrix} \text{(for series circuit element)} \qquad (2.52a)$$

$$\begin{bmatrix} V_1 \\ I_1 \end{bmatrix} = \begin{bmatrix} 1 & 0 \\ Y & 1 \end{bmatrix} \begin{bmatrix} V_2 \\ I_2 \end{bmatrix} \text{(for shunt circuit element)} \qquad (2.52b)$$

Thus, the *ABCD* matrices of a general *T* network shown in Figure 2.8a can be derived as

$$\begin{bmatrix} A & B \\ C & D \end{bmatrix} = \begin{bmatrix} 1 & Z_1 \\ 0 & 1 \end{bmatrix} \begin{bmatrix} 1 & 0 \\ 1/Z_3 & 1 \end{bmatrix} \begin{bmatrix} 1 & Z_2 \\ 0 & 1 \end{bmatrix}$$

$$= \begin{bmatrix} 1 + \dfrac{Z_1}{Z_3} & Z_1 + Z_2 + \dfrac{Z_1 Z_2}{Z_3} \\ \dfrac{1}{Z_3} & 1 + \dfrac{Z_2}{Z_3} \end{bmatrix}. \qquad (2.53)$$

For the π network shown in Figure 2.9a, we have

$$\begin{bmatrix} A & B \\ C & D \end{bmatrix} = \begin{bmatrix} 1 & 0 \\ Y_1 & 1 \end{bmatrix} \begin{bmatrix} 1 & 1/Y_3 \\ 0 & 1 \end{bmatrix} \begin{bmatrix} 1 & 0 \\ Y_2 & 1 \end{bmatrix}$$

$$= \begin{bmatrix} 1 + \dfrac{Y_2}{Y_3} & \dfrac{1}{Y_3} \\ Y_1 + Y_2 + \dfrac{Y_1 Y_2}{Y_3} & 1 + \dfrac{Y_1}{Y_3} \end{bmatrix}. \qquad (2.54)$$

In addition to the above series and shunt lumped elements, the uniform transmission line with a finite length l as shown in Figure 2.15 is a key building element in forming various microwave circuits. According to the relation between voltage and current waves along the line in Equation (2.11), we have

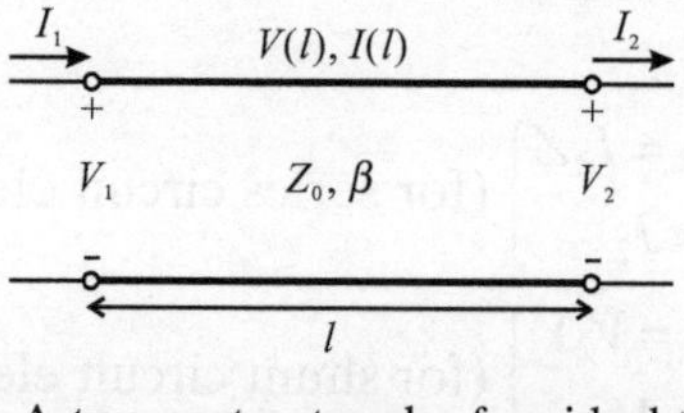

Figure 2.15. A two-port network of an ideal transformer.

$$V(-l) = V_o^+ e^{j\beta l} + V_o^- e^{-j\beta l} \tag{2.55a}$$

$$I(-l) = \frac{V_o^+}{Z_0} e^{j\beta l} - \frac{V_o^-}{Z_0} e^{-j\beta l} \tag{2.55b}$$

$$V(0) = V_o^+ + V_o^- \tag{2.55c}$$

$$I(0) = \frac{V_o^+ - V_o^-}{Z_0}. \tag{2.55d}$$

Substituting Equation (2.55c,d) into Equation (2.55a,b) gives

$$
\begin{aligned}
V(-l) &= \left[\frac{V(0) + I(0)Z_0}{2} \right] e^{j\beta l} + \left[\frac{V(0) - I(0)Z_0}{2} \right] e^{-j\beta l} \\
&= V(0) \left(\frac{e^{j\beta l} + e^{-j\beta l}}{2} \right) + I(0)Z_0 \left(\frac{e^{j\beta l} - e^{-j\beta l}}{2} \right) \\
&= V(0) \cos \beta l + I(0)Z_0 \sin \beta l
\end{aligned}
\tag{2.56a}
$$

$$
\begin{aligned}
I(-l) &= \left[\frac{V(0)/Z_0 + I(0)}{2} \right] e^{j\beta l} - \left[\frac{V(0)/Z_0 - I(0)}{2} \right] e^{-j\beta l} \\
&= \frac{V(0)}{Z_0} \left(\frac{e^{j\beta l} - e^{-j\beta l}}{2} \right) + I(0) \left(\frac{e^{j\beta l} + e^{-j\beta l}}{2} \right) \\
&= V(0) \frac{\sin \beta l}{Z_0} + I(0) \cos \beta l.
\end{aligned}
\tag{2.56b}
$$

Thereafter, the *ABCD* matrix of this finite-length uniform transmission line can be obtained by rearranging Equation (2.56) as

$$
\begin{bmatrix} V(-l) \\ I(-l) \end{bmatrix} = \begin{bmatrix} \cos \beta l & Z_0 \sin \beta l \\ \dfrac{\sin \beta l}{Z_0} & \cos \beta l \end{bmatrix} \begin{bmatrix} V(0) \\ I(0) \end{bmatrix} \tag{2.57}
$$

EXAMPLE 2.3   Derive the *ABCD* parameters of an ideal transformer shown in Figure 2.16.

Solution

For an ideal transformer, the relation of port voltages and currents can be given by the ratio of primary and secondary windings N_1 and N_2 as

$$\frac{V_1}{V_2} = \frac{N_1}{N_2} = \frac{I_2}{I_1}.$$

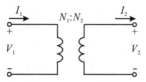

Figure 2.16. A symmetrical two-port network with even- and odd-mode excitations (a) Even-mode excitation. (b) Odd-mode excitation.

Therefore, we can have

$$A = \frac{V_1}{V_2}\bigg|_{I_2=0} = \frac{N_1}{N_2}$$

$$B = \frac{V_1}{I_2}\bigg|_{V_2=0} = 0$$

$$C = \frac{I_1}{V_1}\bigg|_{I_2=0} = 0$$

$$D = \frac{I_1}{I_2}\bigg|_{V_2=0} = \frac{N_2}{N_1}$$

So far, we have discussed and derived the *ABCD* matrices for all the two-port fundamental network elements. As a summary, Table 2.1 lists all the above-derived two-port circuit elements and their *ABCD* matrix. It is the usual case that a filter partially consists of open or short stubs in series or shunt connection with the primary transmission line. Our later discussion on direct synthesis design of UWB bandpass filters starts with such a stub-loaded network. So, the *ABCD* matrices for these stub-loaded transmission lines need to be derived based on the definition of series and shunt circuit elements in Equation (2.52), and they are tabulated in Table 2.2.

2.3.4 Conversions between *S-*, *Z-*, *Y-*, and *ABCD*-Matrix Parameters

- *ABCD*-parameters → *S*-parameters

  Referring to Figure 2.7, the voltage and current at port 2 is given by

$$V_2 = -I_2 Z_{02}, \text{ or } -I_2 = V_2/Z_{02}. \tag{2.58}$$

TABLE 2.1 Some Useful Two-Port Circuit Elements and Their ABCD Matrix

Name	Circuit	ABCD matrix
Series circuit element		$\begin{bmatrix} 1 & Z \\ 0 & 1 \end{bmatrix}$
Shunt circuit element		$\begin{bmatrix} 1 & 0 \\ Y & 1 \end{bmatrix}$
T network		$\begin{bmatrix} 1+\dfrac{Z_1}{Z_3} & Z_1+Z_2+\dfrac{Z_1 Z_2}{Z_3} \\ \dfrac{1}{Z_3} & 1+\dfrac{Z_2}{Z_3} \end{bmatrix}$
π network		$\begin{bmatrix} 1+\dfrac{Y_2}{Y_3} & \dfrac{1}{Y_3} \\ Y_1+Y_2+\dfrac{Y_1 Y_2}{Y_3} & 1+\dfrac{Y_1}{Y_3} \end{bmatrix}$
Line section		$\begin{bmatrix} \cos\beta l & jZ_0 \sin\beta l \\ j\dfrac{\sin\beta l}{Z_0} & \cos\beta l \end{bmatrix}$
Ideal transformer		$\begin{bmatrix} \dfrac{N_1}{N_2} & 0 \\ 0 & \dfrac{N_2}{N_1} \end{bmatrix}$

Noting the change in the sign convention of the current I_2 at port 2, we have

$$V_1 = AV_2 - BI_2 = (A + B/Z_{02})V_2 \qquad (2.59a)$$

$$I_1 = CV_2 - DI_2 = (C + D/Z_{02})V_2. \qquad (2.59b)$$

According to Equation (2.22), the normalized voltages and currents of the forward and backward waves can be derived as a linear function of the port voltage, V_1 or V_2:

$$a_1 = \frac{1}{2}\left(\frac{V_1}{\sqrt{Z_{01}}} + \sqrt{Z_{01}}\,I_1 \right) = \frac{V_2}{2\sqrt{Z_{01}}}(A + B/Z_{02} + CZ_{01} + DZ_{01}/Z_{02}) \quad (2.60a)$$

TABLE 2.2   Some Useful Stub-Load Transmission Line and Their ABCD Matrices

Name	Circuit	ABCD matrix
Shunt open stub		$\begin{bmatrix} 1 & 0 \\ \dfrac{j\tan\beta l}{Z_0} & 1 \end{bmatrix}$
Shunt short stub		$\begin{bmatrix} 1 & 0 \\ -\dfrac{j\cot\beta l}{Z_0} & 1 \end{bmatrix}$
Series open stub		$\begin{bmatrix} 1 & -jZ_0\cot\beta l \\ 0 & 1 \end{bmatrix}$
Series short stub		$\begin{bmatrix} 1 & jZ_0\tan\beta l \\ 0 & 1 \end{bmatrix}$

$$b_1 = \frac{1}{2}\left(\frac{V_1}{\sqrt{Z_{01}}} - \sqrt{Z_{01}}\,I_1\right) = \frac{V_2}{2\sqrt{Z_{01}}}\left(A + B/Z_{02} - CZ_{01} + DZ_{01}/Z_{02}\right) \quad (2.60b)$$

$$b_2 = \frac{1}{2}\left(\frac{V_2}{\sqrt{Z_{02}}} - \sqrt{Z_{02}}\,I_2\right) = \frac{V_2}{\sqrt{Z_{02}}}. \quad (2.60c)$$

Under the restriction of the matching load at port 2, $a_2 = 0$ can be deduced. So, we have

$$S_{11} = \frac{b_1}{a_1}\bigg|_{a_2=0} = \frac{(A + B/Z_{02} - CZ_{01} - DZ_{01}/Z_{02})}{(A + B/Z_{02} + CZ_{01} + DZ_{01}/Z_{02})} \qquad (2.61a)$$

$$S_{21} = \frac{b_2}{a_1}\bigg|_{a_2=0} = \frac{2\sqrt{Z_{01}}/\sqrt{Z_{02}}}{(A + B/Z_{02} + CZ_{01} + DZ_{01}/Z_{02})}. \qquad (2.61b)$$

If port 2 is treated as an input port, the sign conventions of two port currents need to be changed such that

$$\begin{bmatrix} V_2 \\ I_2 \end{bmatrix} = \begin{bmatrix} A & B \\ C & D \end{bmatrix}^{-1} \begin{bmatrix} V_1 \\ -I_1 \end{bmatrix} = \frac{1}{\Delta}\begin{bmatrix} D & B \\ C & A \end{bmatrix}\begin{bmatrix} V_1 \\ -I_1 \end{bmatrix} \qquad (2.62)$$

where Δ is the determinant of the $ABCD$-matrix. If the network is reciprocal, $\Delta = 1$. Replacing the $ABCD$ parameters in (2.61) with the newly defined parameters in (2.62) gives

$$S_{22} = \frac{(D + B/Z_{01} - CZ_{02} - AZ_{02}/Z_{01})}{(D + B/Z_{01} + CZ_{02} + AZ_{02}/Z_{01})} \qquad (2.63a)$$

$$S_{12} = \frac{2\Delta\sqrt{Z_{02}}/\sqrt{Z_{01}}}{(D + B/Z_{01} + CZ_{02} + AZ_{02}/Z_{01})} \qquad (2.63b)$$

- Z-parameters $\rightarrow$ ABCD-parameters
  Based on the definition of Z-parameters in Equation (2.35), we have

$$V_1 = \left(\frac{Z_{11}}{Z_{21}}\right)V_2 + \left(Z_{12} - \frac{Z_{11}Z_{22}}{Z_{21}}\right)I_2 \qquad (2.64a)$$

$$I_1 = \frac{1}{Z_{21}}V_2 - \frac{Z_{22}}{Z_{21}}I_2. \qquad (2.64b)$$

Reversing the sign convention of I_2 in Equation (2.64) for the definition of $ABCD$-parameters gives

$$A = \left(\frac{Z_{11}}{Z_{21}}\right) \quad B = \left(\frac{Z_{11}Z_{22}}{Z_{21}} - Z_{12}\right)$$

$$C = \frac{1}{Z_{21}} \quad D = \frac{Z_{22}}{Z_{21}}. \qquad (2.65)$$

- Z-parameters $\rightarrow$ S-parameters
  Substituting (2.65) into (2.61) and (2.62) gives

$$S_{11} = \frac{(Z_{11} - Z_{01})(Z_{22} + Z_{02}) - Z_{12}Z_{21}}{(Z_{11} + Z_{01})(Z_{22} + Z_{02}) - Z_{12}Z_{21}} \quad S_{12} = \frac{2Z_{12}\sqrt{Z_{01}Z_{02}}}{(Z_{11} + Z_{01})(Z_{22} + Z_{02}) - Z_{12}Z_{21}}$$

$$S_{21} = \frac{2Z_{21}\sqrt{Z_{01}Z_{02}}}{(Z_{11} + Z_{01})(Z_{22} + Z_{02}) - Z_{12}Z_{21}} \quad S_{22} = \frac{(Z_{11} + Z_{01})(Z_{22} - Z_{02}) - Z_{12}Z_{21}}{(Z_{11} + Z_{01})(Z_{22} + Z_{02}) - Z_{12}Z_{21}}.$$

$$(2.66)$$

Similarly, the conversion from Y-parameters to S-parameters can be derived. Table 2.3 provides a detailed list of the conversions or relationship between S-parameters and other network parameters, that is, Y-, Z- and $ABCD$-matrix parameters.

2.4 RELATIVE THEORIES OF NETWORK ANALYSIS

2.4.1 Even- and Odd-Mode Network Analysis

For a two-port symmetrical network shown in Figure 2.17, it usually divides the overall network into two symmetrical bisections. Under even- and odd-mode excitation, the symmetrical plane at the center can be considered as a perfect magnetic wall and electric wall, respectively. In other words, two separated but identical one-port networks can be formed with open- or short-circuited termination. Since a symmetrical two-port network is linear, the port voltages or currents can be simply expressed as the linear sum of those from even and odd excitations. That means the two-port network parameters can be derived from one-port even- and odd-mode bisection networks. Actually, this even- and odd-mode approach is powerful in the analysis of various multi-port symmetrical circuits. For example, analysis of a symmetrical four-port network is given in Reference 5.

Referring to the original two-port network in Figure 2.6 and the rearranged network with the symmetrical plane denoted in Figure 2.17, the port voltages and currents can be expressed as

$$V_1 = V_e + V_o$$
$$I_1 = I_e + I_o$$
$$V_2 = V_e - V_o$$
$$I_2 = I_e - I_o.$$

$$(2.67)$$

TABLE 2.3 Conversion between Two-Port Network Parameters with Arbitrary Terminated Impedances

	S	Z	Y	ABCD
S_{11}	S_{11}	$\dfrac{(Z_{11}-Z_{01})(Z_{22}+Z_{02})-Z_{12}Z_{21}}{(Z_{11}+Z_{01})(Z_{22}+Z_{02})-Z_{12}Z_{21}}$	$\dfrac{(Y_{11}-Y_{01})(Y_{22}+Y_{02})-Y_{12}Y_{21}}{(Y_{11}+Y_{01})(Y_{22}+Y_{02})-Y_{12}Y_{21}}$	$\dfrac{AZ_{02}+B-CZ_{01}Z_{02}-DZ_{01}}{AZ_{02}+B+CZ_{01}Z_{02}+DZ_{01}}$
S_{12}		$\dfrac{2Z_{12}\sqrt{Z_{01}Z_{02}}}{(Z_{11}+Z_{01})(Z_{22}+Z_{02})-Z_{12}Z_{21}}$	$\dfrac{2Y_{12}\sqrt{Y_{01}Y_{02}}}{(Y_{11}+Y_{01})(Y_{22}+Y_{02})-Y_{12}Y_{21}}$	$\dfrac{2(AD-BC)\sqrt{Z_{01}Z_{02}}}{AZ_{02}+B+CZ_{01}Z_{02}+DZ_{01}}$
S_{21}		$\dfrac{2Z_{21}\sqrt{Z_{01}Z_{02}}}{(Z_{11}+Z_{01})(Z_{22}+Z_{02})-Z_{12}Z_{21}}$	$\dfrac{2Y_{21}\sqrt{Y_{01}Y_{02}}}{(Y_{11}+Y_{01})(Y_{22}+Y_{02})-Y_{12}Y_{21}}$	$\dfrac{2\sqrt{Z_{01}Z_{02}}}{AZ_{02}+B+CZ_{01}Z_{02}+DZ_{01}}$
S_{22}		$\dfrac{(Z_{11}+Z_{01})(Z_{22}-Z_{02})-Z_{12}Z_{21}}{(Z_{11}+Z_{01})(Z_{22}+Z_{02})-Z_{12}Z_{21}}$	$\dfrac{(Y_{11}+Y_{01})(Y_{22}-Y_{02})-Y_{12}Y_{21}}{(Y_{11}+Y_{01})(Y_{22}+Y_{02})-Y_{12}Y_{21}}$	$\dfrac{-AZ_{02}+B-CZ_{01}Z_{02}+DZ_{01}}{AZ_{02}+B+CZ_{01}Z_{02}+DZ_{01}}$
Z_{11}	$\dfrac{[(1+S_{11})(1-S_{22})+S_{12}S_{21}]Z_{01}}{(1-S_{11})(1-S_{22})-S_{12}S_{21}}$	Z_{11}	$\dfrac{Y_{22}}{Y_{11}Y_{22}-Y_{12}Y_{21}}$	$\dfrac{A}{C}$
Z_{12}	$\dfrac{2S_{12}\sqrt{Z_{01}Z_{02}}}{(1-S_{11})(1-S_{22})-S_{12}S_{21}}$	Z_{12}	$\dfrac{-Y_{12}}{Y_{11}Y_{22}-Y_{12}Y_{21}}$	$\dfrac{AD-BC}{C}$
Z_{21}	$\dfrac{2S_{21}\sqrt{Z_{01}Z_{02}}}{(1-S_{11})(1-S_{22})-S_{12}S_{21}}$	Z_{21}	$\dfrac{-Y_{21}}{Y_{11}Y_{22}-Y_{12}Y_{21}}$	$\dfrac{1}{C}$
Z_{22}	$\dfrac{[(1-S_{11})(1+S_{22})+S_{12}S_{21}]Z_{02}}{(1-S_{11})(1-S_{22})-S_{12}S_{21}}$	Z_{22}	$\dfrac{Y_{11}}{Y_{11}Y_{22}-Y_{12}Y_{21}}$	$\dfrac{D}{C}$

(Continued)

TABLE 2.3 (Continued)

	S	Z	Y	ABCD
Y_{11}	$\dfrac{[(1-S_{11})(1+S_{22})+S_{12}S_{21}]Y_{01}}{(1+S_{11})(1+S_{22})-S_{12}S_{21}}$	$\dfrac{Z_{22}}{Z_{11}Z_{22}-Z_{12}Z_{21}}$	Y_{11}	$\dfrac{D}{B}$
Y_{12}	$\dfrac{-2S_{12}\sqrt{Y_{01}Y_{02}}}{(1+S_{11})(1+S_{22})-S_{12}S_{21}}$	$\dfrac{-Z_{12}}{Z_{11}Z_{22}-Z_{12}Z_{21}}$	Y_{12}	$\dfrac{BC-AD}{B}$
Y_{21}	$\dfrac{-2S_{21}\sqrt{Y_{01}Y_{02}}}{(1+S_{11})(1+S_{22})-S_{12}S_{21}}$	$\dfrac{-Z_{21}}{Z_{11}Z_{22}-Z_{12}Z_{21}}$	Y_{21}	$\dfrac{-1}{B}$
Y_{22}	$\dfrac{[(1+S_{11})(1-S_{22})+S_{12}S_{21}]Y_{02}}{(1+S_{11})(1+S_{22})-S_{12}S_{21}}$	$\dfrac{Z_{11}}{Z_{11}Z_{22}-Z_{12}Z_{21}}$	Y_{22}	$\dfrac{A}{B}$
A	$\dfrac{[(1+S_{11})(1-S_{22})+S_{12}S_{21}]\sqrt{Z_{01}}}{2S_{12}\sqrt{Z_{02}}}$	$\dfrac{Z_{11}}{Z_{21}}$	$\dfrac{-Y_{22}}{Y_{21}}$	A
B	$\dfrac{[(1+S_{11})(1+S_{22})-S_{12}S_{21}]\sqrt{Z_{01}Z_{02}}}{2S_{12}}$	$\dfrac{Z_{11}Z_{22}-Z_{12}Z_{21}}{Z_{21}}$	$\dfrac{-1}{Y_{21}}$	B
C	$\dfrac{(1-S_{11})(1-S_{22})-S_{12}S_{21}}{2S_{12}\sqrt{Z_{01}Z_{02}}}$	$\dfrac{1}{Z_{21}}$	$\dfrac{-Y_{11}Y_{22}+Y_{12}Y_{21}}{Y_{21}}$	C
D	$\dfrac{[(1-S_{11})(1+S_{22})+S_{12}S_{21}]\sqrt{Z_{02}}}{2S_{12}\sqrt{Z_{01}}}$	$\dfrac{Z_{22}}{Z_{21}}$	$\dfrac{-Y_{11}}{Y_{21}}$	D

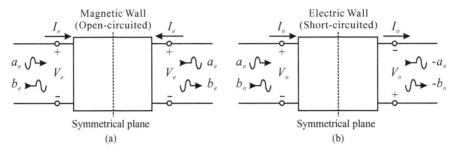

Figure 2.17. Symmetrical T and π networks for even- and odd-mode analysis. (a) T network. (b) π network.

For a one-port network, the even- and odd-mode Z-parameters are given by

$$Z_{11e} = V_e/I_e$$
$$Z_{11o} = V_o/I_o. \tag{2.68}$$

Letting $I_2 = 0$ and substituting Equation (2.68) into Equation (2.67) gives

$$V_1 = Z_{11e}I_e + Z_{11o}I_o$$
$$I_1 = 2I_e = 2I_o \tag{2.69}$$
$$V_2 = Z_{11e}I_e - Z_{11o}I_o.$$

According to the open-circuit conditions in Equation (2.37), we have

$$Z_{11} = \left.\frac{V_1}{I_1}\right|_{I_2=0} = \frac{Z_{11e} + Z_{11o}}{2}$$

$$Z_{21} = \left.\frac{V_2}{I_1}\right|_{I_2=0} = \frac{Z_{11e} - Z_{11o}}{2} \tag{2.70}$$

$$Z_{12} = Z_{21}$$

$$Z_{22} = Z_{11}.$$

Substituting Equation (2.70) into Equation (2.66) gives a set of useful formulas as

$$S_{11} = S_{22} = \frac{Z_{11e}Z_{11o} - Z_0^2}{(Z_{11e} + Z_0)(Z_{11o} + Z_0)} \tag{2.71a}$$

$$S_{21} = S_{12} = \frac{(Z_{11e} - Z_{11o})Z_0}{(Z_{11e} + Z_0)(Z_{11o} + Z_0)} \tag{2.71b}$$

It is of interest to note that Z_{11e} and Z_{11o} also represent the input imped-ances of the one-port even-and odd-mode networks. Thus, we have

$$S_{11} = S_{22} = \frac{Z_{ine}Z_{ino} - Z_0^2}{(Z_{ine} + Z_0)(Z_{ino} + Z_0)} = \frac{Y_0^2 - Y_{ine}Y_{ino}}{(Y_{ine} + Y_0)(Y_{ino} + Y_0)} \quad (2.72a)$$

$$S_{21} = S_{12} = \frac{(Z_{ine} - Z_{ino})Z_0}{(Z_{ine} + Z_0)(Z_{ino} + Z_0)} = \frac{(Y_{ino} - Y_{ine})Y_0}{(Y_{ine} + Y_0)(Y_{ino} + Y_0)}, \quad (2.72b)$$

where $Y_{ine} = 1/Z_{ine}$, $Y_{ino} = 1/Z_{ino}$, $Z_0 = Z_{01} = Z_{02}$, $Y_0 = 1/Z_0$.

EXAMPLE 2.4   Derive the S-parameters of T and π networks in Figure 2.18 using the even- and odd-mode analysis approach.

Solution

For the T network in Figure 2.17a, we can have

$$Z_{ine} = Z_{11e} = Z_a + 2Z_b \quad \text{and} \quad Z_{ino} = Z_{11o} = Z_a.$$

From Equation (2.71) or Equation (2.72),

$$S_{11}^T = S_{22}^T = \frac{(Z_a + 2Z_b)Z_a - Z_0^2}{(Z_a + 2Z_b + Z_0)(Z_a + Z_0)}$$

$$S_{21}^T = S_{12}^T = \frac{(Z_a + 2Z_b - Z_a)Z_0}{(Z_a + 2Z_b + Z_0)(Z_a + Z_0)}.$$

For the π network in Figure 2.17b, we have

$$Y_{ine} = Y_{11e} = Y_a \quad \text{and} \quad Y_{ino} = Y_{11o} = Y_a + 2Y_b.$$

From Equation (2.72),

$$S_{11}^\pi = S_{22}^\pi = \frac{Y_0^2 - Y_a(Y_a + 2Y_b)}{(Y_a + Y_0)(Y_a + 2Y_b + Y_0)}$$

$$S_{21}^\pi = S_{12}^\pi = \frac{(Y_a + 2Y_b - Y_a)Z_0}{(Y_a + Y_0)(Y_a + 2Y_b + Y_0)}.$$

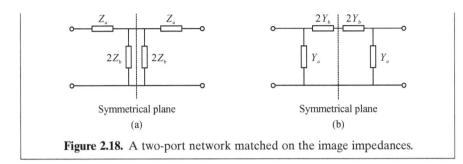

Figure 2.18. A two-port network matched on the image impedances.

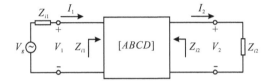

Figure 2.19. A two-port network terminated in a load.

2.4.2   Image Parameter Method

Another useful method for analyzing a two-port network is based on image parameters, for example, image impedance and image propagation constant. It is particularly useful for the design of various matching networks and microwave filters with parallel-coupled lines. The objective of this section is to describe the physical concept and definition of this method, as well as a set of design equations which need to be used in the filter designs in later chapters.

From the guided-wave propagation point of view, the image impedance and image propagation constant of a finite-length circuit network can be somehow considered as two effective parameters per unit length (Z_0 and γ in Eqs. 2.4 and 2.5) of a uniform transmission line, but they are much more powerful than the latter ones in characterizing a two-port network. For a uniform transmission line with length l, the characteristic impedance Z_0 is exactly the same as its image impedance, while the propagation constant multiplied by the electrical length βl is its image propagation constant. It means that the image parameters are more like device parameters and not limited on the analysis of the transmission line. Generally speaking, they can be used not only on symmetrical networks, but also on asymmetrical networks [6].

Now, let us consider a two-port network in Figure 2.19, where the network is specified by *ABCD*-matrix and terminated in its two dissimilar image impedances at two ports. The image impedance at port 1, Z_{i1}, is defined as its input impedance when port 2 is terminated with Z_{i2} (the same definition for Z_{i2}). In the following analysis, our effort is focused to realize perfect impedance matching for the two-port network in Figure 2.19 at both source and load ports with the proper choice of the matched image impedance. Referring to Figure 2.19, the image impedance at the port 1, with $Z_L = Z_{i2}$, is obtained from Equation (2.50) as

$$Z_{i1} = \frac{V_1}{I_1} = \frac{AZ_L + B}{CZ_L + D} = \frac{AZ_{i2} + B}{CZ_{i2} + D}. \tag{2.73}$$

From the inverse $ABCD$-matrix in Equation (2.62), we also have

$$Z_{i2} = \frac{V_2}{-I_2} = \frac{DZ_{i1} + B}{CZ_{i1} + A}. \tag{2.74}$$

Solving Equations (2.73) and (2.74) gives

$$Z_{i1} = \sqrt{\frac{AB}{CD}} \tag{2.75a}$$

$$Z_{i2} = \sqrt{\frac{BD}{AC}}, \tag{2.75b}$$

and

$$AZ_{i2} = DZ_{i1} \tag{2.75c}$$

$$B = CZ_{i1}Z_{i2}. \tag{2.75d}$$

Actually, the image impedances can also be considered as actual impedances of an infinite chain network indicated in Figure 2.20. With proper choice of image impedances for each chained network as done above, the reflection coefficient at each adjunction is expected to become zero. Based on Equation (2.61a,b), we have

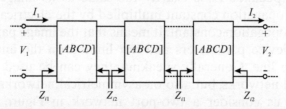

Figure 2.20. A infinite chain networks defined by the image parameters.

$$S_{11} = 0 \Rightarrow AZ_{i2} + B - CZ_{i1}Z_{i2} - DZ_{i1} = 0 \qquad (2.76a)$$

$$S_{22} = 0 \Rightarrow DZ_{i1} + B - CZ_{i1}Z_{i2} - AZ_{i2} = 0, \qquad (2.76b)$$

which yield the same conditions as those in Equation (2.75c,d).

Now, we can introduce the voltage transfer function of a two-port network and define it with reference to Figure 2.19 as

$$\frac{V_2}{V_1} = \sqrt{\frac{Z_{i1}}{Z_{i2}}} e^\gamma, \qquad (2.77)$$

where $\gamma = \alpha + j\beta$ is the image propagation constant. Since $I_2 = V_2/Z_{i2}$ at the load, we have

$$V_1 = AV_2 + BI_2 = AV_2 + \frac{BV_2}{Z_{i2}}. \qquad (2.78)$$

Then, referring to (2.75), (2.77) can be rewritten as

$$\begin{aligned}
e^\gamma &= \sqrt{\frac{Z_{i2}}{Z_{i1}} \frac{V_2}{V_1}} \\
&= \sqrt{\frac{Z_{i2}}{Z_{i1}}} \left(A + \frac{B}{Z_{i2}} \right) \\
&= \sqrt{\frac{D}{A}} \left(A + B\sqrt{\frac{AC}{BD}} \right) \\
&= \sqrt{AD} + \sqrt{BC}.
\end{aligned} \qquad (2.79)$$

Since the reciprocal network is satisfied with the condition in Equation (2.49), that is, $AD - BC = 1$, we can have

$$e^{-\gamma} = \frac{1}{e^\gamma} = \frac{1}{\sqrt{AD} + \sqrt{BC}} = \frac{\sqrt{AD} - \sqrt{BC}}{AD - BC} = \sqrt{AD} - \sqrt{BC}. \quad (2.80)$$

Thus, the image propagation constant can be expressed as

$$\cosh \gamma = \sqrt{AD} \qquad (2.81a)$$

or

$$\gamma = \cosh^{-1}\left(\sqrt{AD}\right). \qquad (2.81b)$$

The relation between the image parameters and Z-/Y-matrix parameters of a general two-port network can be established, and it is tabulated in Table 2.4.

TABLE 2.4   Conversion between Image Parameters and Other General Circuit Parameters

	ABCD	Z	Y	Z&Y
Z_{i1}	$\sqrt{\dfrac{AB}{CD}}$	$\sqrt{\dfrac{Z_{11}\lvert Z\rvert}{Z_{22}}}$	$\sqrt{\dfrac{Y_{22}\lvert Y\rvert}{Y_{11}}}$	$\sqrt{\dfrac{Z_{11}}{Y_{11}}}$
Z_{i2}	$\sqrt{\dfrac{BD}{AC}}$	$\sqrt{\dfrac{Z_{22}\lvert Z\rvert}{Z_{11}}}$	$\sqrt{\dfrac{Y_{11}\lvert Y\rvert}{Y_{22}}}$	$\sqrt{\dfrac{Z_{22}}{Y_{22}}}$
γ	$\cosh^{-1}\left(\sqrt{AD}\right)$ $\sinh^{-1}\left(\sqrt{BC}\right)$ $\coth^{-1}\left(\sqrt{\dfrac{AD}{BC}}\right)$	$\cosh^{-1}\left(\dfrac{\sqrt{Z_{11}Z_{22}}}{Z_{21}}\right)$ $\sinh^{-1}\left(\dfrac{\sqrt{\lvert Z\rvert}}{Z_{21}}\right)$ $\coth^{-1}\left(\sqrt{\dfrac{Z_{11}Z_{22}}{\lvert Z\rvert}}\right)$	$\cosh^{-1}\left(\dfrac{\sqrt{Y_{11}Y_{22}}}{Y_{21}}\right)$ $\sinh^{-1}\left(\dfrac{\sqrt{\lvert Y\rvert}}{Y_{21}}\right)$ $\coth^{-1}\left(\sqrt{\dfrac{Y_{11}Y_{22}}{\lvert Y\rvert}}\right)$	$\coth^{-1}\left(\sqrt{Z_{11}Y_{11}}\right)$ $\coth^{-1}\left(\sqrt{Z_{22}Y_{22}}\right)$

$$A=\sqrt{\frac{Z_{i1}}{Z_{i2}}}\cosh\gamma \qquad Z_{11}=Z_{i1}\coth\gamma \qquad Y_{11}=Y_{i1}\coth\gamma$$

$$B=\sqrt{Z_{i1}Z_{i2}}\sinh\gamma \qquad Z_{12}=Z_{21}=\frac{\sqrt{Z_{i1}Z_{i2}}}{\sinh\gamma} \qquad Y_{12}=Y_{21}=\frac{-\sqrt{Z_{i1}Z_{i2}}}{\sinh\gamma}$$

$$C=\frac{\sinh\gamma}{\sqrt{Z_{i1}Z_{i2}}} \qquad Z_{22}=Z_{i2}\coth\gamma \qquad Y_{22}=Y_{i2}\coth\gamma$$

$$D=\sqrt{\frac{Z_{i2}}{Z_{i1}}}\cosh\gamma$$

where $\lvert Z\rvert = Z_{11}Z_{22}-Z_{12}Z_{21}$, $\lvert Y\rvert = Y_{11}Y_{22}-Y_{12}Y_{21}$

EXAMPLE 2.5   Derive the image parameters of L, T, and π networks shown in Figure 2.21.

Solution

Referring to Equation (2.52), $ABCD$-matrix of L network shown in Figure 2.21a is obtained as

$$\begin{bmatrix} A & B \\ C & D \end{bmatrix}^{L}=\begin{bmatrix} 1 & Z_a \\ 0 & 1 \end{bmatrix}\begin{bmatrix} 1 & 0 \\ 1/Z_b & 1 \end{bmatrix}=\begin{bmatrix} 1+Z_a/Z_b & Z_a \\ 1/Z_b & 1 \end{bmatrix}.$$

Based on the relations in Table 2.4, we have

$$Z_{i1}^{L}=\sqrt{\frac{AB}{CD}}=\sqrt{Z_a Z_b}\sqrt{1+Z_a/Z_b}$$

$$Z_{i2}^L = \sqrt{\frac{BD}{AC}} = \frac{\sqrt{Z_a Z_b}}{\sqrt{1 + Z_a/Z_b}}$$

$$\gamma^L = \cosh^{-1}\left(\sqrt{AD}\right) = \cosh^{-1}\sqrt{1 + Z_a/Z_b}.$$

From Table 2.1, the $ABCD$-matrix of a symmetrical T network shown in Figure 2.21b is given by

$$\begin{bmatrix} 1 + \dfrac{Z_a}{Z_b} & 2Z_a + \dfrac{Z_a^2}{Z_b} \\ \dfrac{1}{Z_b} & 1 + \dfrac{Z_a}{Z_b} \end{bmatrix}.$$

Then,

$$Z_{i1}^T = Z_{i2}^T = \sqrt{\frac{B}{C}} = \sqrt{Z_a(Z_a + 2Z_b)}$$

$$\gamma^T = \cosh^{-1}\left(\sqrt{AD}\right) = \cosh^{-1}(1 + Z_a/Z_b).$$

In a similar way, the image parameters of the π network shown in Figure 2.21c is given in its admittances as

$$Y_{i1}^\pi = Y_{i2}^\pi = \sqrt{Y_a(Y_a + 2Y_b)}$$

$$\gamma^\pi = \cosh^{-1}(1 + Y_a/Y_b).$$

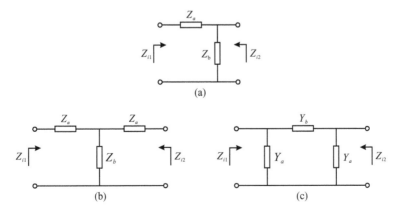

(a)

(b)   (c)

Figure 2.21. Symmetrical T and π networks for even- and odd-mode analysis. (a) T network. (b) π network.

2.5 SUMMARY

The fundamental theory for transmission lines and analysis methodology for microwave networks have been presented in this chapter to provide a theoretical basis in filter analysis that will be done in the later chapters. In particular, we have discussed various basic transmission line models and characterized transmission lines with varied loaded impedances. Then, different types of network parameters, such as Z-, Y-, S-, $ABCD$-matrix, have been introduced, and their relationships have been further revealed. Meanwhile, the even- and odd-mode analysis approach has been introduced via analysis of a symmetrical two-port network. Finally, the image parameters have been introduced for analysis of a general two-port network with symmetrical or asymmetrical configurations, and they are very useful in analysis of various chain networks.

REFERENCES

1 R. E. Collin, *Foundations for Microwave Engineering*, 2nd ed., McGraw-Hill, New York, 1992.
2 D. M. Pozar, *Microwave Engineering*, 2nd ed., John Wiley & Sons, Inc., New York, 1998.
3 J.-S. Hong and M. J. Lancaster, *Microstrip Filters for RF/Microwave Applications*, John Wiley & Sons, Inc., New York, 2001.
4 H.-R. Ahn, *Asymmetric Passive Components in Microwave Integrated Circuits*, John Wiley & Sons, Inc., Hoboken, NJ, 2006.
5 J. Reed and G. J. Weeler, "A method of analysis of symmetrical four-port networks," *IRE Trans. Microw. Theory Tech.* 4 (1956) 346–352.
6 G. Matthaei, L. Young, and E. M. T. Jones, *Microwave Filters, Impedance-Matching Network, and Coupled Structures*, Artech House, Dedham, MA, 1980.

CHAPTER 3

CONVENTIONAL PARALLEL-COUPLED LINE FILTER

3.1 INTRODUCTION

Parallel-coupled line bandpass filter was first proposed by Cohn in 1958 [1] and has been considered as one of the most useful bandpass filters over the decades. Nowadays, although numerous advanced filter design techniques have been presented by the researchers, parallel-coupled line filter is still one of the most important filter topologies in many practical applications due to its matured design procedure and well-documented design parameters. In recent years, full-wave simulation tools were successfully developed to allow the automatic analysis and design of the parallel-coupled line filters with the user-defined specifications. In this chapter, the design procedure for the parallel-coupled line filters will be reviewed. In order to show a full picture of the filter development, we would like to start with the classic lumped-element lowpass filter prototype. After that, lowpass to bandpass transformation will be introduced. Finally, the lumped bandpass network is linked to the conventional parallel-coupled line implementation. The design formulas are all explicitly derived step-by-step to clearly describe a systematic design procedure.

Microwave Bandpass Filters for Wideband Communications, First Edition. Lei Zhu, Sheng Sun, Rui Li.
© 2012 John Wiley & Sons, Inc. Published 2012 by John Wiley & Sons, Inc.

3.2 LUMPED-ELEMENT LOWPASS FILTER PROTOTYPE

Most of the conventional lowpass, highpass, bandpass, and bandstop microwave filters are evolved and designed from an initial lowpass prototype filter through the standard procedure of transformation, impedance, and frequency scaling. The normalized element values of the lowpass prototype filter are derived from various network synthesis methods [1–4] and tabulated for convenience of filter designers [5,6]. The basic idea of the network synthesis method or insertion loss method starts with the finding and formulation of a closed-form transfer function that can exactly present the frequency response of a filter. Maximally flat and Chebyshev equal-ripple functions are two commonly used transfer functions.

3.2.1 Maximally Flat and Chebyshev Characteristics

For a lowpass filter with maximally flat or Butterworth response, its power loss ratio or insertion loss can be expressed as the following polynomial function in the unit of dB

$$L_A(\omega) = 10\log_{10}\left[1+\varepsilon^2\left(\frac{\omega}{\omega_c}\right)^{2n}\right], \tag{3.1a}$$

and

$$\varepsilon = \sqrt{10^{\frac{L_{Ar}}{10}}-1}, \tag{3.1b}$$

where ω_c is the cutoff frequency and n is the order of a filter. The band edge is defined at the cutoff frequency, in which the attenuation is denoted as L_{Ar}. If $L_{Ar} = 3.0$ dB is selected as usual, we have $\varepsilon = 1$. The response, named as the maximally flat one, refers that the quantity in the bracket of the logarithm function, which in fact is the power transfer function, has $(2n - 1)$ zero derivatives at $\omega = 0$. Figure 3.1 shows the frequency responses of a typical maximally flat lowpass filter with the order of $n = 3$ and 6, respectively.

For an n-order lowpass filter with Chebyshev or equal ripple response, its insertion loss function can be obtained as

$$L_A(\omega) = 10\log_{10}\left[1+\varepsilon^2 T_n^2\left(\frac{\omega}{\omega_c}\right)\right], \tag{3.2a}$$

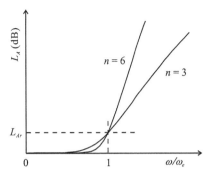

Figure 3.1. Maximally flat lowpass filter response ($n = 3$ and 6).

and

$$\varepsilon = \sqrt{10^{\frac{L_{Ar}}{10}} - 1}, \qquad (3.2b)$$

where $T_n(x)$ is the Chebyshev polynomial of the first kind with a degree of n. The first few Chebyshev polynomial functions are given

$$T_1(x) = x, \qquad (3.3a)$$

$$T_2(x) = 2x^2 - 1, \qquad (3.3b)$$

$$T_3(x) = 4x^3 - 3x, \qquad (3.3c)$$

$$T_4(x) = 8x^4 - 8x^2 + 1, \qquad (3.3d)$$

$$T_5(x) = 16x^5 - 20x^3 + 5x. \qquad (3.3e)$$

A higher-order polynomial can be deduced in terms of two lower-order polynomials through the recurrence formula as

$$T_n(x) = 2xT_{n-1}(x) - T_{n-2}(x). \qquad (3.4)$$

The Chebyshev polynomial of the first kind also has the following trigonometric identity

$$T_n(\cos\theta) = \cos n\theta. \qquad (3.5)$$

Hence, the insertion loss function of the Chebyshev lowpass filter becomes

$$L_A(\omega) = 10\log_{10}\left\{1 + \varepsilon^2 \cos^2\left[n\cos^{-1}\left(\frac{\omega}{\omega_c}\right)\right]\right\}, \quad \text{for } |\omega| \le \omega_c, \quad (3.6a)$$

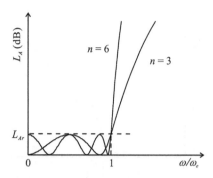

Figure 3.2. Frequency responses of a Chebyshev lowpass filter ($n = 3$ and 6).

and

$$L_A(\omega) = 10\log_{10}\left\{1 + \varepsilon^2\cosh^2\left[n\cosh^{-1}\left(\frac{\omega}{\omega_c}\right)\right]\right\}, \text{ for } |\omega| \geq \omega_c. \quad (3.6b)$$

The frequency responses of a Chebyshev lowpass filter with $n = 3$ and 6 are plotted in Figure 3.2 with the same scale as Figure 3.1 for comparative analysis. It is observed that the Chebyshev filter has sharper roll-off at cutoff frequency as compared with the maximally flat filter with the same order. However, the maximally flat response has less insertion loss and less delay distortion in the passband. There are also other types of transfer functions, such as elliptic function and linear phase function. There are always trade-offs among different kinds of transfer functions. It is very difficult to have a "perfect" transfer function with all the desired responses. The selection of transfer function depends on the feature of the applications. If shaper skirt near cutoff frequencies is required, Chebyshev and elliptic functions become two better candidates. However, if group delay is more concerned, maximally flat and linear phase characteristics are preferred.

3.2.2 Lumped-Element Ladder Network

The lumped-element lowpass filter prototype and its dual network with identical response are shown in Figure 3.3. For these lowpass filter prototypes, the element values are normalized so that the source impedance is equal to 1 and the cutoff frequency becomes $\omega_c = 1$. The element values, $g_0, g_1, g_2, \ldots, g_n$, and g_{n+1}, indicated in Figure 3.3, are sequentially numbered from the generator impedance g_0 to the load impedance g_{n+1}. There are n reactive elements in-between, and they are shunt and series

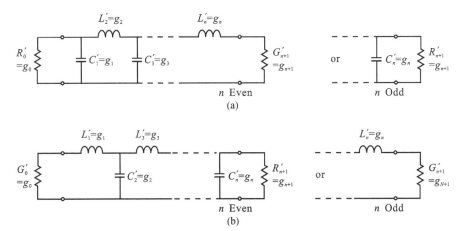

Figure 3.3. Lumped-element ladder networks for lowpass filter prototypes. (a) Prototype with a shunt capacitor as the first element. (b) Prototype with a series inductance as the first element.

elements alternately. The network may begin with either a shunt element or a series element as shown in Figure 3.3a,b. The last reactive element before the load can be either in shunt or series connection, depending on whether n is an odd or even integer number. The element values indicated in Figure 3.3 are defined as follows:

$$g_0 = \begin{cases} \text{generator resistance } R_0, \text{ if } g_1 = C_1' \text{ (network in Figure 3.3a)} \\ \text{generator conductance } G_0, \text{ if } g_1 = L_1' \text{ (network in Figure 3.3b)} \end{cases}$$

$$\underset{k=1 \text{ to } n}{g_k} = \begin{cases} \text{inductance of series inductors} \\ \text{capacitance of shunt capacitors} \end{cases}$$

$$g_{n+1} = \begin{cases} \text{load resistance } R_{n+1}, \text{ if } g_n = C_n' \\ \text{load conductance } G_{n+1}, \text{ if } g_n = L_n' \end{cases}.$$

In order to achieve the exact frequency responses specified by the transfer functions in Section 3.2.1, the element values in the filter networks shown in Figure 3.3 have to be determined. In the following, we will show the detailed procedure on how to derive those element values for the two types of transfer functions discussed above.

3.2.2.1 Maximally Flat Lowpass Prototype

A two-element lowpass filter prototype starting with a series element is shown in Figure 3.4. We assume that both the source and load impedances are 1, and

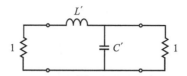

Figure 3.4. Lowpass filter prototype ($n = 2$).

the cutoff frequency $\omega_c = 1$. For this two-port network with both the generator and the load matched, the *reflection coefficient* Γ (S_{11}) and *transmission coefficient* T (S_{21}) can be derived in terms of the *ABCD* matrix parameters from Equation (2.61) as

$$\Gamma = \frac{(A-D)+(B-C)}{A+B+C+D}, \tag{3.7a}$$

$$T = \frac{2}{A+B+C+D}. \tag{3.7b}$$

For a lossless network as we discussed here, we have

$$|\Gamma|^2 + |T|^2 = 1. \tag{3.8}$$

Therefore, substituting Equation (3.8) into Equation (2.32) gives the power transfer function as

$$\frac{P_{\text{Avail}}}{P_L} = \frac{1}{|T|^2} = \frac{|T|^2 + |\Gamma|^2}{|T|^2} = 1 + \left|\frac{\Gamma}{T}\right|^2 = 1 + \left|\frac{(A-D)+(B-C)}{2}\right|^2. \tag{3.9}$$

The *ABCD* matrix of the overall circuit is

$$\begin{bmatrix} A & B \\ C & D \end{bmatrix} = \begin{bmatrix} 1 & j\omega L' \\ 0 & 1 \end{bmatrix} \begin{bmatrix} 1 & 0 \\ j\omega C' & 1 \end{bmatrix}$$

$$= \begin{bmatrix} 1 - \omega^2 L'C' & j\omega L' \\ j\omega C' & 1 \end{bmatrix}. \tag{3.10}$$

So, the power transfer function Equation (3.9) becomes

$$\frac{P_{\text{Avail}}}{P_L} = 1 + \left|\frac{-\omega^2 L'C' + j(\omega L' - \omega C')}{2}\right|^2$$

$$= 1 + \frac{1}{4}[(L'^2 + C'^2 - 2L'C')\omega^2 + L'^2 C'^2 \omega^4], \tag{3.11}$$

From Equation (3.1a), the insertion loss function for the maximally flat characteristic response with $n = 2$ and $\omega_c = 1$ is simplified as

$$\frac{P_{\text{Avail}}}{P_L} = 1 + \omega^4. \tag{3.12}$$

To equalize Equation (3.11) with Equation (3.12), the coefficients of different orders of ω in these two equations must be exactly the same. Hence, the coefficient of ω^2 in Equation (3.11) must equal to zero due to the absence of ω^2 term in Equation (3.12), such that

$$\begin{cases} \dfrac{1}{4}(L'^2 + C'^2 - 2L'C') = 0 \\ \dfrac{1}{4}L'^2C'^2 = 1 \end{cases}. \tag{3.13}$$

From Equation (3.13), we can determine the two unknown reactive element values as

$$L' = C' = \sqrt{2}.$$

This analytical method works well for calculating the element values of maximally flat lowpass prototype. In particular, this procedure can be further simplified when the network is symmetrical $(A = D)$, where the power transfer function in Equation (3.9) becomes

$$\frac{P_{\text{Avail}}}{P_L} = 1 + \left| \frac{B - C}{2} \right|^2. \tag{3.14}$$

For a maximally flat lowpass filter with $L_{Ar} = 3$ dB and $\omega_c = 1$, a series of closed-form formulas are given in Equation (3.15). Then, all the element values can be determined, and they are tabulated in Table 3.1 for $n = 1$ to 10.

$$g_0 = 1,$$

$$g_k = 2\sin\left[\frac{(2k-1)\pi}{2n}\right], \qquad k = 1, 2, \ldots, n, \tag{3.15}$$

$$g_{n+1} = 1.$$

TABLE 3.1 Element Values for a Maximally Flat Low-Pass Filter Prototype ($g_0 = 0$, $\omega_c = 1$, $n = 1\text{-}10$)

n	g_1	g_2	g_3	g_4	g_5	g_6	g_7	g_8	g_9	g_{10}	g_{11}
1	2.000	1.000									
2	1.414	1.414	1.000								
3	1.000	2.000	1.000	1.000							
4	0.7654	1.848	1.848	0.7654	1.000						
5	0.6180	1.618	2.000	1.618	0.6180	1.000					
6	0.5176	1.414	1.932	1.932	1.414	0.5176	1.000				
7	0.4450	1.247	1.802	2.000	1.802	1.247	0.4450	1.000			
8	0.3902	1.111	1.663	1.962	1.962	1.663	1.111	0.3902	1.000		
9	0.3473	1.000	1.532	1.879	2.000	1.879	1.532	1.000	0.3473	1.000	
10	0.3129	0.9080	1.414	1.782	1.975	1.975	1.782	1.414	0.9080	0.3129	1.000

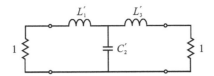

Figure 3.5. Lowpass filter prototype ($n = 3$).

3.2.2.2 Chebyshev Equal-Ripple Lowpass Filter Prototype

Figure 3.5 shows a lowpass prototype with odd $n = 3$. Both the generator and the load resistance are normalized to unity or 1. The $ABCD$ matrix of the entire circuit is

$$\begin{bmatrix} A & B \\ C & D \end{bmatrix} = \begin{bmatrix} 1 & j\omega L_1' \\ 0 & 1 \end{bmatrix}\begin{bmatrix} 1 & 0 \\ j\omega C_2' & 1 \end{bmatrix}\begin{bmatrix} 1 & j\omega L_3' \\ 0 & 1 \end{bmatrix}$$

$$= \begin{bmatrix} 1-\omega^2 L_1' C_2' & j(\omega L_1' + \omega L_3' - \omega^3 L_1' L_3' C_2') \\ j\omega C_2' & 1-\omega^2 L_2' C_2' \end{bmatrix}. \tag{3.16}$$

Using Equation (3.14), the power transfer function is

$$\frac{P_{\text{Avail}}}{P_L} = 1 + \left| \frac{(C_2' L_3' - L_1' C_2')\omega^2 + j(\omega L_1' - \omega C_2' + \omega L_3' - \omega^3 L'_1 L_3' C_2')}{2} \right|^2$$

$$= 1 + \frac{1}{4}\Big[(L_1'^2 - 2L_1' C_2' + C_2'^2 + 2L_1' L_3' - 2L_3' C_2' + L_3'^2)\omega^2$$

$$+ (L_1'^2 C_2'^2 - 2L_1'^2 C_2' L_3' - 2L_1' C_2' L_3'^2 + L_3'^2 C_2'^2)\omega^4 + L_1'^2 L_3'^2 C_2'^2 \omega^6 \Big]. \tag{3.17}$$

The desired response with $L_{Ar} = 0.5$ dB and $n = 3$ is obtained from Equation (3.2a) as

$$\frac{P_{\text{Avail}}}{P_L} = 1 + \varepsilon^2 (4\omega^3 - 3\omega)^2$$

$$= 1 + \varepsilon^2 (9\omega^2 - 24\omega^4 + 16\omega^6), \tag{3.18}$$

where $\varepsilon^2 = 10^{0.05} - 1$. By equalizing Equation (3.17) with Equation (3.18), we have the following equations:

$$\begin{cases} \dfrac{1}{4}(L_1'^2 - 2L_1'C_2' + C_2'^2 + 2L_1'L_3' - 2L_3'C_2' + L_3'^2) = 9\varepsilon^2 \\ \dfrac{1}{4}(L_1'^2C_2'^2 - 2L_1'^2C_2'L_3' - 2L_1'C_2'L_3'^2 + L_3'^2C_2'^2) = -24\varepsilon^2 \qquad (3.19) \\ \dfrac{1}{4}L_1'^2L_3'^2C_2'^2 = 16\varepsilon^2 \end{cases}$$

Solving Equation (3.19), we have

$$L_1' = L_3' = 1.5963,$$

$$C_2' = 1.0967.$$

Thus, it can be concluded that the ladder network discussed above, with odd integers of n, is definitely symmetrical. However, for even integers of n, this network becomes unsymmetrical and antimetrical [6] about its middle. In this situation, one half of the network is the reciprocal of the other half with respect to a positive real constant R_h, and it is defined as

$$R_h = \sqrt{R_0 R_{n+1}}, \qquad (3.20)$$

where R_0 and R_{n+1} are the generator and the load resistances, as shown in Figure 3.3a. If Z_k is the impedance of one reactive element in one half of the ladder network, its dual branch at the other half of the network Z_{n+1-k} becomes

$$Z_{n+1-k} = \frac{R_h^2}{Z_k}. \qquad (3.21)$$

Hence, for a Chebyshev lowpass prototype filter with even n, the element values of the second half of the network can be obtained from the element values of the first half as

$$C_{n+1-k}' = \frac{L_k'}{R_h^2}, \qquad (3.22a)$$

$$L_{n+1-k}' = R_h^2 C_k'. \qquad (3.22b)$$

When n becomes large, it is quite tedious to calculate these element values by oneself. For a Chebyshev lowpass prototype filter with pass-band ripple level L_{Ar}, cutoff frequency $\omega_c = 1$, and normalized generator impedance $g_0 = 1$, the following closed formulas can be formed to determine those element values as tabulated in Table 3.2.

TABLE 3.2 Element values for Chebyshev Low-Pass Filter Prototypes ($g_0 = 0$, $\omega_c = 1$, $L_{Ar} = 0.01, 0.1, 1.0$ dB)

n	g_1	g_2	g_3	g_4	g_5	g_6	g_7	g_8	g_9	g_{10}	g_{11}
0.01 dB ripple											
1	0.0960	1.0000									
2	0.4488	0.4077	1.1007								
3	0.6291	0.9702	0.6291	1.000							
4	0.7128	1.2003	1.3212	0.6476	1.1007						
5	0.7563	1.3049	1.5773	1.3049	0.7563	1.000					
6	0.7813	1.3600	1.6896	1.5350	1.4970	0.7098	1.1007				
7	0.7969	1.3924	1.7481	1.6331	1.7481	1.3924	0.7969	1.000			
8	0.8072	1.4130	1.7824	1.6833	1.8529	1.6193	1.5554	0.7333	1.1007		
9	0.8144	1.4270	1.8043	1.7125	1.9057	1.7125	1.8043	1.4270	0.8144	1.000	
10	0.8196	1.4369	1.8192	1.7311	1.9362	1.7590	1.9055	1.6527	1.5817	0.7446	1.1007
0.1 dB ripple											
1	0.3052	1.0000									
2	0.8430	0.6220	1.3554								
3	1.0315	1.1474	1.0315	1.0000							
4	1.1088	1.3061	1.7703	0.8180	1.3554						
5	1.1468	1.3712	1.9750	1.3712	1.1468	1.0000					
6	1.1681	1.4039	2.0562	1.5170	1.9029	0.8618	1.3554				
7	1.1811	1.4228	2.0966	1.5733	2.0966	1.4228	1.1811	1.0000			
8	1.1897	1.4346	2.1199	1.6010	2.1699	1.5640	1.9444	0.8778	1.3554		
9	1.1956	1.4425	2.1345	1.6167	2.2053	1.6167	2.1345	1.4425	1.1956	1.0000	
10	1.1999	1.4481	2.1444	1.6265	2.2253	1.6418	2.2046	1.5821	1.9628	0.8853	1.3554

(*Continued*)

TABLE 3.2 (*Continued*)

n	g_1	g_2	g_3	g_4	g_5	g_6	g_7	g_8	g_9	g_{10}	g_{11}
1.0 dB ripple											
1	1.0177	1.0000									
2	1.8219	0.6850	2.6599								
3	2.0236	0.9941	2.0236	1.0000							
4	2.0991	1.0644	2.8311	0.7892	2.6599						
5	2.1349	1.0911	3.0009	1.0911	2.1349	1.0000					
6	2.1546	1.1041	3.0634	1.1518	2.9367	0.8101	2.6599				
7	2.1664	1.1116	3.0934	1.1736	3.0934	1.1116	2.1664	1.0000			
8	2.1744	1.1161	3.1107	1.1839	3.1488	1.1696	2.9685	0.8175	2.6599		
9	2.1797	1.1192	3.1215	1.1897	3.1747	1.1897	3.1215	1.1192	2.1797	1.0000	
10	2.1836	1.1213	3.1286	1.1933	3.1890	1.1990	3.1738	1.1763	2.9824	0.8210	2.6599

$$g_1 = \frac{2a_1}{\sinh\left(\dfrac{\beta}{2n}\right)},$$

$$g_k = \frac{4a_{k-1}a_k}{b_{k-1}g_{k-1}} \qquad k = 1, 2, 3, \ldots, n, \tag{3.23}$$

$$g_k = 1 \qquad\qquad \text{for } n \text{ odd}$$

$$= \coth^2\left(\frac{\beta}{4}\right) \qquad \text{for } n \text{ even},$$

where

$$\beta = \ln\left(\coth\frac{L_{Ar}}{17.37}\right),$$

$$a_k = \sin\left[\frac{(2k-1)\pi}{2n}\right] \qquad k = 1, 2, 3, \ldots, n,$$

$$b_k = \sinh^2\left(\frac{\beta}{2n}\right) + \sin^2\left(\frac{k\pi}{n}\right) \qquad k = 1, 2, 3, \ldots, n.$$

3.3 IMPEDANCE AND FREQUENCY TRANSFORMATION

In Section 3.2, a lowpass prototype filter was introduced, where all the element values involved are normalized with respect to the source impedance/conductance. In practice, other types of filters rather than lowpass filters need to be designed at different operating frequency bands. Without repeating the entire derivation procedure, highpass, bandpass, and bandstop filters with required termination values and cutoff or center frequencies can be derived by virtue of impedance and frequency transformations or mapping techniques from the initial lowpass prototype filter. The impedance transformation can be done by scaling the normalized generator impedance or conductance to a desired Z_0 or Y_0. The scaling factor γ_0 is defined as

$$\gamma_0 = \begin{cases} Z_0/g_0 & \text{for } g_0 \text{ being the resistance} \\ g_0/Y_0 & \text{for } g_0 \text{ being the conductance.} \end{cases} \tag{3.24}$$

Without loss of generality, the following impedance scaling can be applied to characterize the other reactive or resistive elements shown in Figure 3.3:

$$L \to \gamma_0 L, \qquad (3.25a)$$

$$C \to C/\gamma_0, \qquad (3.25b)$$

$$R \to \gamma_0 R, \qquad (3.25c)$$

$$G \to G/\gamma_0. \qquad (3.25d)$$

3.3.1 Lowpass to Lowpass Transformation

To transform a lowpass prototype filter with unity cutoff frequency to a lowpass filter with cutoff frequency ω_c, the frequency variable has to be scaled by a factor of $1/\omega_c$, or the cutoff frequency 1 has to be mapped to a realistic cutoff frequency ω_c. The transformation procedure is achieved by replacing ω by ω/ω_c, which is

$$\omega \to \frac{\omega}{\omega_c}, \qquad (3.26)$$

After the impedance and frequency transformation are executed, the reactive elements in Figure 3.3 become

$$L_k = \frac{g_k \gamma_0}{\omega_c}, \qquad (3.27a)$$

$$C_k = \frac{g_k}{\omega_c \gamma_0}. \qquad (3.27b)$$

3.3.2 Lowpass to Highpass Transformation

For a lowpass to highpass frequency transformation, the substitution of the frequency dependent is as follows:

$$\omega \to -\frac{\omega_c}{\omega}, \qquad (3.28)$$

The transformation actually swaps the passband and stopband in the lowpass prototype to the stopband and passband in a highpass prototype filter. Looking at the individual reactive elements, it is found that the lowpass network with series inductors and shunt capacitors are now transformed to a highpass network with series capacitors and shunt inductors. Thus, the element values in the transformed filter networks are

$$L_k = \frac{\gamma_0}{\omega_c g_k} \qquad \text{for } g_k \text{ being capacitance} \qquad (3.29a)$$

$$C_k = \frac{1}{\omega_c g_k \gamma_0} \qquad \text{for } g_k \text{ being inductance.} \qquad (3.29b)$$

3.3.3 Lowpass to Bandpass Transformation

The lowpass to bandpass transformation is a little bit complicated. Instead of a single cutoff frequency for lowpass and highpass, two frequencies ω_1 and ω_2 are used to denote the lower and upper passband edges. The transformation is done by replacing ω with a term as

$$\omega \rightarrow \frac{1}{FBW}\left(\frac{\omega}{\omega_0} - \frac{\omega_0}{\omega}\right), \qquad (3.30)$$

where FBW stands for *fractional bandwidth*, and it is defined as

$$FBW = \frac{\omega_2 - \omega_1}{\omega_0}, \qquad (3.31)$$

and ω_0 is the center frequency of the transformed bandpass filter, and it is defined as the geometric mean of ω_1 and ω_2 as

$$\omega_0 = \sqrt{\omega_1 \omega_2}. \qquad (3.32)$$

As a result, a series element in the lowpass prototype filter is transformed to a series resonant circuit, whereas a shunt element is converted to a parallel resonant circuit. In the analysis, both types of series and parallel resonant circuits have the same resonant frequency of ω_0. Therefore, the element values in the series resonators can be determined as

$$L_k = \frac{g_k \gamma_0}{FBW \omega_0}, \qquad (3.33a)$$

$$C_k = \frac{FBW}{\omega_0 g_k \gamma_0}. \qquad (3.33b)$$

Similarly, the element values in the parallel resonators can be derived as

$$L_k = \frac{FBW\gamma_0}{\omega_0 g_k},$$ (3.34a)

$$C_k = \frac{g_k}{FBW\omega_0\gamma_0}.$$ (3.34b)

3.3.4   Lowpass to Bandstop Transformation

The lowpass to bandstop transformation can be achieved in a similar way as the lowpass to bandpass transformation. The angular frequency ω is mapped in such a way that

$$\omega \rightarrow -FBW\left(\frac{\omega}{\omega_0} - \frac{\omega_0}{\omega}\right)^{-1},$$ (3.35)

where FBW and ω_0 have the same definition as Equations (3.31) and (3.32). Under this transformation, a series element in the lowpass prototype is transformed to a parallel resonant circuit, which is opposite to that in the lowpass to bandpass transformation, such that

$$L_k = \frac{FBWg_k\gamma_0}{\omega_0},$$ (3.36a)

$$C_k = \frac{1}{FBW\omega_0 g_k\gamma_0}.$$ (3.36b)

Meanwhile, a shunt element is converted to a series resonant circuit as

$$L_k = \frac{\gamma_0}{FBW\omega_0 g_k},$$ (3.37a)

$$C_k = \frac{FBWg_k}{\omega_0\gamma_0}.$$ (3.37b)

Figure 3.6 summaries the element transformation from a lowpass prototype filter to highpass, bandpass and bandstop filters. The values of L_k and C_k in Figure 3.6a,e can be found from Equations (3.27), (3.29), (3.33), (3.34), and (3.36), (3.37).

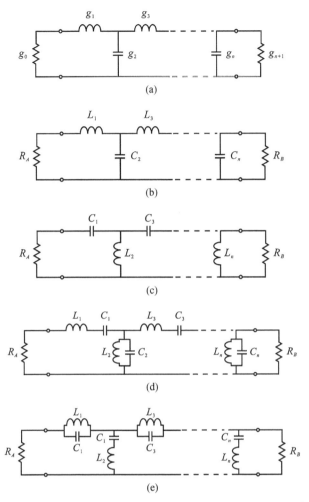

Figure 3.6. Lumped-element ladder network of lowpass prototype and transformed networks. (a) Lowpass prototype. (b) Transformed lowpass filter. (c) Transformed highpass filter. (d) Transformed bandpass filter. (e) Transformed bandstop filter.

EXAMPLE 3.1 Bandpass filter design.

Design a three-pole Chebyshev bandpass filter with a center frequency at 2 GHz, a *FBW* of 10%, and an in-band ripple level of 0.1 dB. The impedances at the two ports are 50 Ω.

Solution

The element values for a lowpass prototype with $L_{Ar} = 0.1$ dB are found in Table 3.2 as

$$g_0 = g_4 = 1 \quad g_1 = g_3 = 1.0315 \quad g_2 = 1.1474.$$

By selecting the impedance scaling factor as $\gamma_0 = 50$, the transformed element values can be calculated from Equations (3.33) and (3.34) for a prototype lowpass filter starting with a series element, as shown in Figure 3.7.

$$L_1 = \frac{g_1 \gamma_0}{FBW\omega_0} = 41.042 \text{ nH} \quad C_1 = \frac{FBW}{\omega_0 g_1 \gamma_0} = 0.154 \text{ pF}$$

$$L_2 = \frac{FBW\gamma_0}{\omega_0 g_2} = 0.347 \text{ nH} \quad C_2 = \frac{g_2}{FBW\omega_0\gamma_0} = 18.261 \text{ pF}$$

$$L_1 = \frac{g_1 \gamma_0}{FBW\omega_0} = 41.042 \text{ nH} \quad C_1 = \frac{FBW}{\omega_0 g_1 \gamma_0} = 0.154 \text{ pF}$$

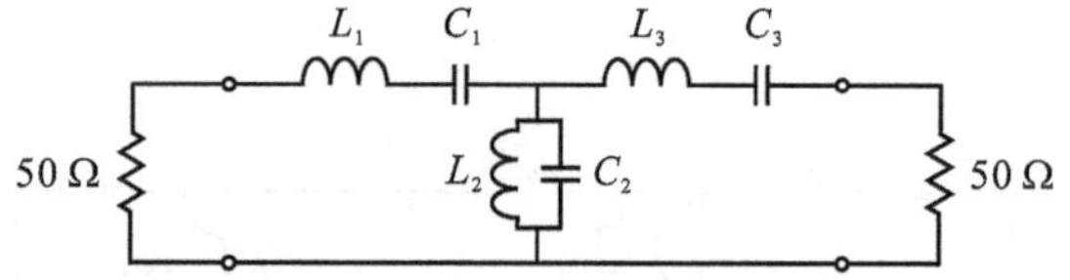

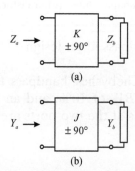

Figure 3.7. Bandpass filter in Example 3.1.

3.4 IMMITTANCE INVERTERS

Immittance inverter is classified as impedance inverter and admittance inverter. An ideal impedance inverter works as a quarter-wavelength impedance transformer with A characteristic impedance of K at all frequencies. Figure 3.8a shows the two-port impedance inverter with

Figure 3.8. Definition and block diagram of immittance inverters. (a) Impedance inverter. (b) Admittance inverter.

one port terminated by a load Z_b, the impedance looking from the other port is

$$Z_a = \frac{K^2}{Z_b}. \tag{3.38}$$

The admittance inverter operates in a similar way. For an ideal admittance inverter, it has the same property as a quarter-wavelength transformer with characteristic admittance of J at all frequencies. Therefore, if one end of the admittance inverter is terminated by an admittance Y_b as shown in Figure 3.8b, the admittance Y_a looking from the other end is

$$Y_a = \frac{Y^2}{Y_b}. \tag{3.39}$$

As the most attractive feature, the immittance inverter is useful in converting a series capacitor/inductor to a shunt inductor/capacitor with a phase shift of $\pm 90°$ or an odd multiple thereof and vice versa. The $ABCD$ matrix of the ideal impedance and admittance inverters are

$$\begin{bmatrix} A & B \\ C & D \end{bmatrix} = \begin{bmatrix} 0 & \pm jK \\ \mp \dfrac{1}{jK} & 0 \end{bmatrix}, \tag{3.40a}$$

and

$$\begin{bmatrix} A & B \\ C & D \end{bmatrix} = \begin{bmatrix} 0 & \pm \dfrac{1}{jJ} \\ \mp jJ & 0 \end{bmatrix}. \tag{3.40b}$$

3.5 LOWPASS PROTOTYPE FILTER WITH IMMITTANCE INVERTER

The ladder network of a lowpass prototype filter discussed in Section 3.2.2 is composed of series inductances cascading with shunt capacitances in an alternate arrangement. Making use of the unique property of immittance inverters, the lowpass prototype in Figure 3.3 can be converted to a filter network shown in Figure 3.9. The converted network consists of either series inductors or shunt capacitors that are connected via immittance inverters. Since the immittance inverters can shift the impedance or admittance levels depending on the values of K and J parameters, the new element values L_{ai} and C_{ai}, termination

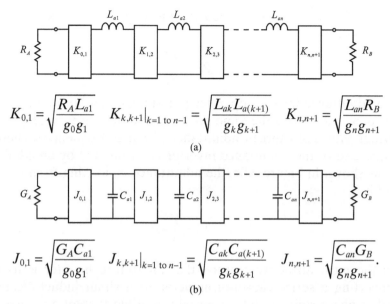

$$K_{0,1} = \sqrt{\frac{R_A L_{a1}}{g_0 g_1}} \qquad K_{k,k+1}\big|_{k=1 \text{ to } n-1} = \sqrt{\frac{L_{ak} L_{a(k+1)}}{g_k g_{k+1}}} \qquad K_{n,n+1} = \sqrt{\frac{L_{an} R_B}{g_n g_{n+1}}}$$

(a)

$$J_{0,1} = \sqrt{\frac{G_A C_{a1}}{g_0 g_1}} \qquad J_{k,k+1}\big|_{k=1 \text{ to } n-1} = \sqrt{\frac{C_{ak} C_{a(k+1)}}{g_k g_{k+1}}} \qquad J_{n,n+1} = \sqrt{\frac{C_{an} G_B}{g_n g_{n+1}}}.$$

(b)

Figure 3.9. Lowpass prototype filter with (a) impedance inverter, (b) admittance inverter.

resistance R_A and R_B, as well as the termination conductance G_A and G_B, can be arbitrarily chosen. All the formulas on $K_{i,i+1}$ and $J_{i,i+1}$ listed in Figure 3.3 have to be satisfied in order to have the same frequency response as the prototype filter in Figure 3.3. The g_i parameters in Figure 3.9 are the element values that were derived above for an n-order lowpass prototype filter.

Now, let us illustrate the procedure to derive the parameters of $K_{k,k+1}$ and $J_{k,k+1}$ in Figure 3.9. Figure 3.10a shows one portion of the lowpass prototype filter with a series element L'_k and a shunt element C'_k. By scaling the impedance level of the network by a factor of L_{ak}/L'_k, the network in Figure 3.10a is in a form as Figure 3.10b, where Y_k and Y_{k+1} indicate the admittance looking from the left and right of the shunt element. Next, an impedance inverter is employed to convert the shunt element to a series element in Figure 3.10c. In order to keep the filter performance undistorted, Y_k in both cases should be identical such that

$$Y_{k+1} + j\omega\left(\frac{L_k}{L_{ak}}\right)C_{k+1} = \frac{1}{K_{k,k+1}^2}\left(Z'_{k+1} + j\omega L_{a(k+1)}\right). \qquad (3.41)$$

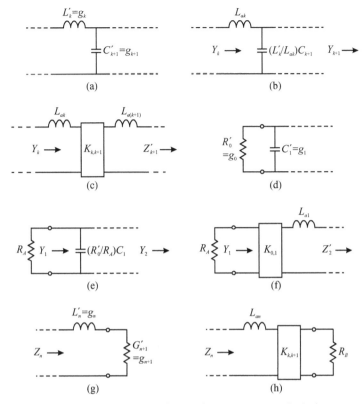

Figure 3.10. Transformation and K parameter calculation.

By equalizing the coefficients of frequency variable ω, we have

$$K_{k,k+1} = \sqrt{\frac{L_{ak} L_{a(k+1)}}{g_n g_{n+1}}}.$$ (3.42)

In addition, we have the relationship between Y_{k+1} and Z'_{k+1} as

$$Y_{k+1} = \frac{Z'_{k+1}}{K^2_{k,k+1}}.$$ (3.43)

It can be verified that Y_{k+1} is zero if an open circuit is assumed beyond the capacitor C'_{k+1}. Similarly, due to existence of a short circuit after the conversion in Figure 3.10c, the value for Z'_{k+1} is 0 and Y_{k+1} becomes zero from Equation (3.43) as well. The first and last K in the network can be derived in a similar manner. Figure 3.10d shows the first active element and the normalized generator impedance in the original low-pass prototype. As the impedance level is lifted up to R_A, an impedance scaling needs to be carried out such that the shunt capacitor is transformed to a series inductor by $K_{0,1}$. Again, the input admittance Y_1 in Figure 3.10e,f should equal to each other in order to keep the filter response unchanged, leading to

$$j\omega\left(\frac{R_0}{R_A}\right)C_1 + Y_2 = \frac{j\omega L_{a1} + Z_2'}{K_{0,1}^2}. \tag{3.44}$$

By indentifying the terms with same dependent variable ω, $K_{0,1}$ can be derived as

$$K_{0,1} = \sqrt{\frac{R_A L_{a1}}{g_0 g_1}}. \tag{3.45}$$

For the last inversion network, the impedance level of the network is shifted by L_{an}/L_n ($=R_B/G_{n+1}$). After introducing the K inverter, the input impedances Z_n should be scaled by a factor of L_{an}/L_n, so we have

$$\left(\frac{L_{an}}{L_n}\right)\left(j\omega L_n + \frac{1}{G_{n+1}}\right) = \frac{K_{n,n+1}^2}{R_B} + j\omega L_{an}. \tag{3.46}$$

Applying the same method, $K_{n,n+1}$ can be generally expressed as

$$K_{n,n+1} = \sqrt{\frac{L_{an} R_B}{g_n g_{n+1}}}. \tag{3.47}$$

With a similar conversion procedure, the lowpass prototype can be converted to a network that consists of only shunt elements connected via admittance inverters, as shown in Figure 3.9b. The lowpass filter with immittance inverter can then be transformed to a bandpass or other type of filters in a way that we introduced in Section 3.3. The transformed bandpass filter and its dual with immittance inverter, K or J, are shown in Figure 3.11.

These two generalized networks are composed of series or shunt lumped resonators and impedance or admittance inverters. When the frequency increases to the microwave regime, the lumped components are difficult to be fabricated or realized as expected in theory. Instead, we need to use distributed elements to make up either equivalent series or shunt resonators. In practice, these series or shunt LC resonators can be constructed in various forms, such as transmission line resonators or waveguide resonators. In this context, *resonant frequency* ω_0 and *slope parameter* are usually used to characterize such a resonator regardless of their types. For a series resonator with reactance $X(\omega)$, the reactance slope parameter is defined as

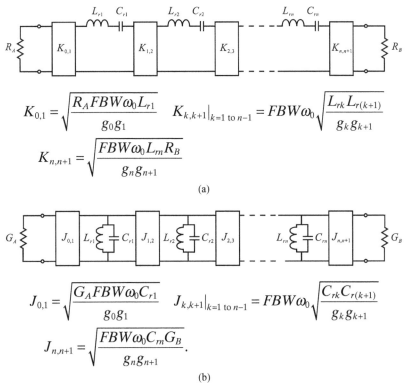

$$K_{0,1} = \sqrt{\frac{R_A FBW \omega_0 L_{r1}}{g_0 g_1}} \qquad K_{k,k+1}\big|_{k=1 \text{ to } n-1} = FBW \omega_0 \sqrt{\frac{L_{rk} L_{r(k+1)}}{g_k g_{k+1}}}$$

$$K_{n,n+1} = \sqrt{\frac{FBW \omega_0 L_{rn} R_B}{g_n g_{n+1}}}$$

(a)

$$J_{0,1} = \sqrt{\frac{G_A FBW \omega_0 C_{r1}}{g_0 g_1}} \qquad J_{k,k+1}\big|_{k=1 \text{ to } n-1} = FBW \omega_0 \sqrt{\frac{C_{rk} C_{r(k+1)}}{g_k g_{k+1}}}$$

$$J_{n,n+1} = \sqrt{\frac{FBW \omega_0 C_{rn} G_B}{g_n g_{n+1}}}.$$

(b)

Figure 3.11. Bandpass filter with (a) impedance inverters, (b) admittance inverters.

$$x = \frac{\omega_0}{2} \frac{dX(\omega)}{d\omega}\bigg|_{\omega=\omega_0} \text{ ohms.} \tag{3.48}$$

For a shunt resonator with susceptance $B(\omega)$, the susceptance slope parameter is

$$b = \frac{\omega_0}{2} \frac{dB(\omega)}{d\omega}\bigg|_{\omega=\omega_0} \text{ mhos.} \tag{3.49}$$

Under the resonant condition, the series or shunt LC resonators, shown in Figure 3.11, between the immittance inverters must have zero reactance or susceptance at ω_0, so that

$$L_{rk} C_{rk} = \frac{1}{\omega_0^2}. \tag{3.50}$$

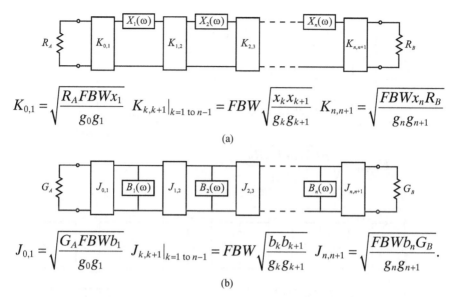

$$K_{0,1} = \sqrt{\frac{R_A FBW x_1}{g_0 g_1}} \quad K_{k,k+1}\big|_{k=1 \text{ to } n-1} = FBW\sqrt{\frac{x_k x_{k+1}}{g_k g_{k+1}}} \quad K_{n,n+1} = \sqrt{\frac{FBW x_n R_B}{g_n g_{n+1}}}$$

(a)

$$J_{0,1} = \sqrt{\frac{G_A FBW b_1}{g_0 g_1}} \quad J_{k,k+1}\big|_{k=1 \text{ to } n-1} = FBW\sqrt{\frac{b_k b_{k+1}}{g_k g_{k+1}}} \quad J_{n,n+1} = \sqrt{\frac{FBW b_n G_B}{g_n g_{n+1}}}.$$

(b)

Figure 3.12. Generalized bandpass filters with (a) impedance inverter, (b) admittance inverter.

The reactance and susceptance slope parameters are

$$x_k = \omega_0 L_{rk} = \frac{1}{\omega_0 C_{rk}}, \tag{3.51a}$$

$$b_k = \omega_0 C_{rk} = \frac{1}{\omega_0 L_{rk}}. \tag{3.51b}$$

By replacing $\omega_0 L_{rk}$ and $\omega_0 C_{rk}$ in the two sets of design equations in Figure 3.11 with the slope parameters in Equation (3.51) while generalizing the series and shunt resonators with the specified reactance and susceptance, we can build up a generalized bandpass filter network and its dual network, as shown in Figure 3.12.

3.6 PARALLEL-COUPLED LINE BANDPASS FILTER

One popular form to implement the bandpass filter network in Figure 3.11 is to use the cascaded parallel-coupled line resonators as shown in Figure 3.13. In the following, we will verify that equivalent circuits of these two filter networks are the same as each other at the center frequency of operation. As the basic element in this filter, the parallel-coupled line section shown in Figure 3.14a can be equivalently modeled

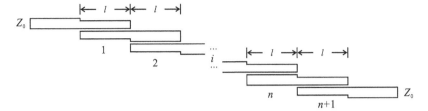

Figure 3.13. Parallel-coupled transmission line bandpass filter.

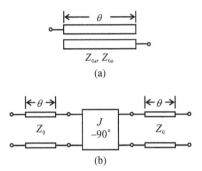

Figure 3.14. Parallel-coupled line and its equivalent circuit.

as a *J*-inverter with two sections of transmission lines shown in Figure 3.14b.

To verify such equivalence of these two networks, their overall *ABCD* matrices should be deduced. From Reference 7, the *ABCD* matrix of a parallel-coupled line in Figure 3.14a is derived using its even- and odd-mode characteristic impedances, Z_{0e} and Z_{0o}, as

$$\begin{bmatrix} A & B \\ C & D \end{bmatrix}_a = \begin{bmatrix} \dfrac{Z_{0e}+Z_{0o}}{Z_{0e}-Z_{0o}}\cos\theta & j\dfrac{(Z_{0e}-Z_{0o})^2-(Z_{0e}+Z_{0o})^2\cos^2\theta}{2(Z_{0e}-Z_{0o})\sin\theta} \\ j\dfrac{2\sin\theta}{Z_{0e}-Z_{0o}} & \dfrac{Z_{0e}+Z_{0o}}{Z_{0e}-Z_{0o}}\cos\theta \end{bmatrix}. \quad (3.52)$$

From Table 2.1, the *ABCD* matrix of a transmission line section with electrical length of θ and characteristic impedance Z_0 is found as

$$\begin{bmatrix} \cos\theta & jZ_0\sin\theta \\ \dfrac{j\sin\theta}{Z_0} & \cos\theta \end{bmatrix}. \quad (3.53)$$

Throughout the discussion in Chapter 3, we can understand that the
J-inverter in Figure 3.14b actually indicates an ideal admittance inverter
with characteristic admittance J and the phase shift of $-90°$ at all frequencies. Thus, the $ABCD$ matrix of the J-inverter is expressed as

$$\begin{bmatrix} 0 & \dfrac{-j}{J} \\ -jJ & 0 \end{bmatrix}. \tag{3.54}$$

Hence, the $ABCD$ matrix of the entire J-inverter circuit in Figure 3.14b
can be obtained by cascading the $ABCD$ matrices of three element
blocks as

$$
\begin{aligned}
\begin{bmatrix} A & B \\ C & D \end{bmatrix}_b &= \begin{bmatrix} \cos\theta & jZ_0\sin\theta \\ \dfrac{j\sin\theta}{Z_0} & \cos\theta \end{bmatrix} \begin{bmatrix} 0 & \dfrac{-j}{J} \\ -jJ & 0 \end{bmatrix} \begin{bmatrix} \cos\theta & jZ_0\sin\theta \\ \dfrac{j\sin\theta}{Z_0} & \cos\theta \end{bmatrix} \\
&= \begin{bmatrix} \left(JZ_0 + \dfrac{1}{JZ_0}\right)\sin\theta\cos\theta & j\left(JZ_0^2\sin^2\theta - \dfrac{\cos^2\theta}{J}\right) \\ j\left(\dfrac{1}{JZ_0^2}\sin^2\theta - J\cos^2\theta\right) & \left(JZ_0 + \dfrac{1}{JZ_0}\right)\sin\theta\cos\theta \end{bmatrix}.
\end{aligned}
\tag{3.55}
$$

Under the exact equalization of the two resultant matrices in Equations
(3.52) and (3.55), we can derive the following equations:

$$\frac{Z_{0e} + Z_{0o}}{Z_{0e} - Z_{0o}}\cos\theta = \left(JZ_0 + \frac{1}{JZ_0}\right)\sin\theta\cos\theta, \tag{3.56a}$$

$$\frac{(Z_{0e} - Z_{0o})^2 - (Z_{0e} + Z_{0o})^2\cos^2\theta}{2(Z_{0e} - Z_{0o})\sin\theta} = JZ_0^2\sin^2\theta - \frac{\cos^2\theta}{J}, \tag{3.56b}$$

$$\frac{2\sin\theta}{Z_{0e} - Z_{0o}} = \frac{1}{JZ_0^2}\sin^2\theta - J\cos^2\theta. \tag{3.56c}$$

By solving Equation (3.56a,c), the even- and odd-mode impedances of
the parallel-coupled lines with arbitrary length can be determined as
follows:

$$\frac{Z_{0e}}{Z_0} = \frac{1 + JZ_0\csc\theta + J^2Z_0^2}{1 - J^2Z_0^2\cot^2\theta}, \tag{3.57a}$$

$$\frac{Z_{0o}}{Z_0} = \frac{1 - JZ_0\csc\theta + J^2Z_0^2}{1 - J^2Z_0^2\cot^2\theta}. \tag{3.57b}$$

In the vicinity of $\theta = \pi/2$, where $\sin\theta \approx 1$ and $\cos\theta \approx 0$, the above equations are simplified as

$$Z_{0e} = Z_0(1 + JZ_0 + J^2 Z_0^2), \tag{3.58a}$$

$$Z_{0o} = Z_0(1 - JZ_0 + J^2 Z_0^2), \tag{3.58b}$$

or

$$Z_{0e} = \frac{1}{Y_0}\left(1 + \frac{J}{Y_0} + \left(\frac{J}{Y_0}\right)^2\right), \tag{3.59a}$$

$$Z_{0e} = \frac{1}{Y_0}\left(1 - \frac{J}{Y_0} + \left(\frac{J}{Y_0}\right)^2\right). \tag{3.59b}$$

Based on the above analysis, the parallel-coupled line can be only reasonably modeled as an admittance inverter with two transmission lines in a small frequency range near the center frequency, and it may be very approximate for a wide frequency bandwidth.

When a few sections of parallel-coupled lines are cascaded as illustrated in Figure 3.13, the overall network in Figure 3.15a can be obtained by cascading the equivalent networks in Figure 3.14b, and it

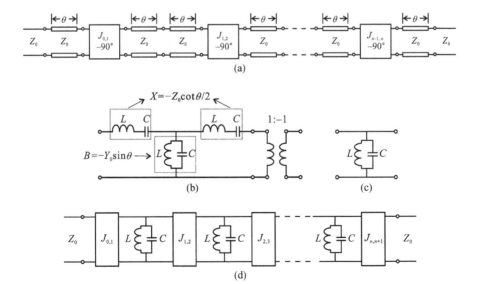

Figure 3.15. Equivalent circuit of a parallel-coupled bandpass filter.

is composed of a few transmission line sections with electrical length of 2θ or $\lambda_g/2$ at the center frequency via J-inverters between two neighboring networks. As shown in Figure 3.15b, an exact equivalent lumped element model of the transmission line with length of 2θ consists of a T-network and an ideal transformer. Since the transformer is only used to reverse a relevant phase, it does not affect the insertion loss response of the filter. The reactance parameters of two series LC resonators are equal to $X = -Z_0\cot\theta/2$, then they become negligibly small when θ is in the vicinity of π. Similarly, the susceptance B is very small near $\theta = \pi$, but the impedance of such a shunt component is very large so as to be dominant in the whole network. Therefore, the circuit in Figure 3.15b is approximately equivalent to that of Figure 3.15c. By replacing the transmission line sections in Figure 3.15a with this lumped model, we can yield the circuit in Figure 3.15d.

Now, we need to determine the values of L and C in Figure 3.15c in terms of the parameters of the transmission line sections. Since the susceptance of the shunt component is $B = -Y_0\sin\theta$ near $\theta = \pi$, we can get

$$B = -Y_0(\pi - \theta) = \pi Y_0\left(\frac{\lambda_{g0}}{\lambda_g} - 1\right), \tag{3.60}$$

where λ_g and λ_{g0} are the guided wavelength and guided wavelength at the center frequency. With reference to Figure 3.15c, the susceptance of the lumped model in Figure 3.15c is derived as

$$B = \omega C - \frac{1}{\omega L}. \tag{3.61}$$

Under the equal condition of the susceptance slope parameters for two circuits, we have

$$2C = \frac{\pi Y_0}{\omega_0}. \tag{3.62}$$

So, the lumped capacitor can be calculated as

$$C = \frac{\pi}{2Z_0\omega_0}, \tag{3.63}$$

By virtue of the resonant condition, we have

$$LC = \frac{1}{\omega_0^2}. \tag{3.64}$$

The lumped inductor can be accordingly obtained as

$$L = \frac{2Z_0}{\pi\omega_0}. \tag{3.65}$$

By substituting Equation (3.64) into the equations in Figure 3.11b or Figure 3.12b, the design equations for a bandpass filter with $n + 1$ sections can be finally obtained as follows:

$$\frac{J_{0,1}}{Y_0} = \sqrt{\frac{\pi FBW}{2g_0 g_1}}, \tag{3.66a}$$

$$\frac{J_{k,k+1}}{Y_0} = \frac{\pi FBW}{2} \frac{1}{\sqrt{g_k g_{k+1}}} \qquad \text{for } k = 2, 3, \ldots, n. \tag{3.66b}$$

$$\frac{J_{n,n+1}}{Y_0} = \sqrt{\frac{\pi FBW}{2g_n g_{n+1}}}. \tag{3.66c}$$

By substituting Equation (3.65a–c) into Equation (3.59), the even- and odd-mode characteristic impedances of each coupled line section can be then calculated.

The bandpass filter using parallel-coupled line with open ends shown in Figure 3.13 has its dual structure, which is parallel-coupled line with short-circuited ends, as shown in Figure 3.16 [2]. The even- and odd-mode characteristic admittances of the parallel-coupled sections can be obtained from the even- and odd-mode characteristic impedances in (3.58) with their mutual relationship as follows:

$$(Y_{0o})_{k,k+1} = Y_0^2 (Z_{0e})_{k,k+1}, \tag{3.67a}$$

$$(Y_{0e})_{k,k+1} = Y_0^2 (Z_{0o})_{k,k+1}. \tag{3.67b}$$

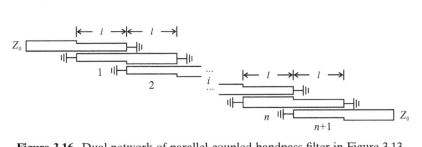

Figure 3.16. Dual network of parallel-coupled bandpass filter in Figure 3.13.

EXAMPLE 3.2   Parallel-coupled line bandpass filter design.

Design two five-pole coupled line bandpass filters with (a) open ends or (b) short ends. The ripple level in the passband is 0.1 dB, the center frequency is 2 GHz, and the *FBW* is 10%. The characteristic impedances of the input and output ports are $Z_0 = 50\ \Omega$.

Solution

The element values of a Chebyshev lowpass prototype with $L_{Ar} = 0.1$ dB and $n = 5$ can be found out from Table 3.2 as

$$g_0 = g_6 = 1 \quad g_1 = g_5 = 1.1468$$
$$g_2 = g_4 = 1.3712 \quad g_3 = 1.9750.$$

(a)   Bandpass filter with coupled lines ended with open circuits:
By substituting the above values into Equations (3.66) and (3.59) with $FBW = 10\%$ and $Z_0 = 50\ \Omega$, we can calculate the admittance parameter, even- and odd-mode impedances as tabulated as

n	$J_{n,n+1}/Y_0$	$Z_{0e}\ (\Omega)$	$Z_{0o}\ (\Omega)$
0 and 5	0.3701	75.35	38.34
1 and 4	0.1253	57.05	44.52
2 and 3	0.0955	55.23	45.68

(b)   Bandpass filter with coupled lines ended with short circuits:
The relevant parameters can be similarly calculated with the values tabulated as

n	$Y_{0e}\ (\Omega)$	$Y_{0o}\ (\Omega)$
0 and 5	65.21	33.18
1 and 4	56.15	43.82
2 and 3	54.73	45.27

The lengths of all the parallel-coupled sections are set as one quarter-wavelength at the center frequency of 2.0 GHz. Regardless of different topologies, the two above-designed filters have achieved exactly identical frequency responses as shown in Figure 3.17.

The parallel-coupled line bandpass filter has been implemented on various transmission line structures. Microstrip line is one of the most popular structures used for realizing this type of filter, but there are a few disadvantages coming along with microstrip parallel-coupled line bandpass filter. Due to the non-TEM property of a microstrip parallel-coupled line on the inhomogeneous substrate, the phase velocities for even- and odd-modes become dissimilar, and it results to distortion of the upper-stopband performance. To circumvent this most critical problem, several techniques, such as substrate overlays and overcoupled resonators, have been developed. In addition, the open-end fringing effect increases the electrical length of a line. This discontinuity can be compensated by reducing the line length by an amount of Δl, and the explicit formula has been provided in Reference 8. The bandwidth of parallel-coupled line filter can be enlarged by increasing the order of the filter, but only up to 20% because the line /gap widths can hardly be reduced to an extreme extent. In Reference 9, the bandwidth of a parallel-coupled line filter is extended above 45% with two quarter-wavelength transformers at the input and output ports.

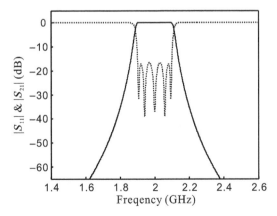

Figure 3.17. Simulated frequency response of the parallel-coupled line bandpass filter of Example 3.2.

3.7 SUMMARY

This chapter reviewed the synthesis design of the conventional band-pass filters based on the topology of a parallel-coupled transmission line. Starting with the lumped-element lowpass filter prototype, the transformations between different filtering types have been introduced. After the immittance inverter is defined and derived, the design formulas for the parallel-coupled lines are given for synthesis design of bandpass filters with defined specifications.

REFERENCES

1 S. B. Cohn, "Parallel-coupled transmission-line-resonator filters," *IRE Trans. Microw. Theory Tech.* MTT-6 (1958) 223–231.

2 G. L. Matthaei, "Design of wide band (and narrow band) bandpass microwave filters on the insertion loss basis," *IRE Trans. Microw. Theory Tech.* MTT-8 (1960) 580–593.

3 E. G. Cristal, "New design equations for a class of microwave filters," *IEEE Trans. Microw. Theory Tech.* MTT-19 (1971) 486–490.

4 M. C. Horton and R. J. Wenzel, "General theory and design of optimum quarter-wave TEM filters," *IEEE Trans. Microw. Theory Tech.* MTT-13(3) (1965) 316–327.

5 A. I. Zverev, *Handbook of Filter Synthesis*, Wiley, New York, 1967.

6 G. Matthaei, L. Young, and E. M. T. Jones, *Microwave Filters, Impedance-Matching Network, and Coupled Structures*, Artech House, Dedham, MA, 1980.

7 G. I. Zysman and A. K. Johnson, "Coupled transmission line networks in an inhomogeneous dielectric medium," *IEEE Trans. Microw. Theory Tech.* MTT-17(10) (1969) 753–759.

8 K. C. Gupta, R. Garg, I. Bahl, and P. Bhartia, *Microstrip Lines and Slotlines*, 2nd ed., Artech House, Norwood, MA, 1996.

9 P. A. Kirton and K. K. Pang, "Extending the realizable bandwidth of edge-coupled stripline filters," *IEEE Trans. Microw. Theory Tech.* MTT-25 (1977) 672–676.

CHAPTER 4

PLANAR TRANSMISSION LINE RESONATORS

4.1 INTRODUCTION

Microwave resonators have been widely used in many applications, such as filters, oscillators, amplifiers, and microwave measurement. Among different types of resonators, the planar transmission line resonator has been considered as the most commonly used resonator for application in radio frequency (RF) and microwave integrated circuits. Unlike waveguide or dielectric resonators, which can offer very high quality factor (Q-factor), planar transmission line resonators, constructed with microstrip line, coplanar waveguide (CPW), and coplanar-stripline (CPS) topology, usually have ordinary Q values due to their intrinsic loss factors of metallization, dielectric substrate, and radiation. However, because of their attractive features, such as low-cost, compact size, lightweight, easy massive fabrication, and so on, a large number of integrated filters have been developed using planar resonators for various communication systems [1].

Figure 4.1 shows the cross-sectional view of a few commonly used planar transmission lines, including stripline, finline, microstrip line, CPW, slotline, and CPS. By terminating a finite portion of a uniform

Microwave Bandpass Filters for Wideband Communications, First Edition. Lei Zhu, Sheng Sun, Rui Li.
© 2012 John Wiley & Sons, Inc. Published 2012 by John Wiley & Sons, Inc.

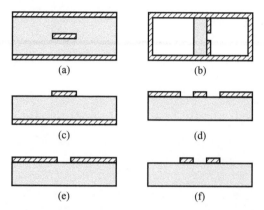

Figure 4.1. Cross-sectional views of some commonly used transmission lines. (a) Stripline. (b) Finline. (c) Microstrip line. (d) Coplanar waveguide (CPW). (e) Slotline. (f) Coplanar stripline (CPS).

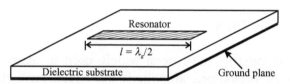

Figure 4.2. Three-dimensional geometry of a half-wavelength microstrip open-circuited resonator.

transmission line with either open-circuits or short-circuits at two ends, a transmission line resonator can be made up. Figure 4.2 shows a half-wavelength microstrip open-circuited resonator whose two terminals are open ended. Over the past decades, this resonator has been widely used in the design of parallel-coupled line filters by virtue of the well-developed synthesis approach. Such a uniform impedance resonator (UIR) resonates periodically at integer multiples of its fundamental resonant frequency, which could be harmful for the single-band application, but useful for the multiple-band and wideband applications. In this chapter, resonant mode behaviors of the UIR and stepped impedance resonator (SIR) with nonuniform configuration will be discussed and analyzed in terms of transmission line theory. Our particular focus is to demonstrate the ratios of higher-order resonances to the fundamental one of these resonators and to adjust them for the design of single-, multiple-, or wide-band bandpass filters.

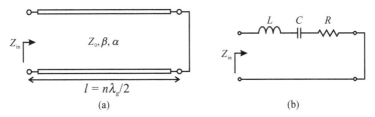

Figure 4.3. Resonant circuit model for a short-circuit lossy transmission line resonator with length l. (a) Transmission line resonant circuit. (b) Its equivalent series type of lumped-element resonant circuit.

4.2   UNIFORM IMPEDANCE RESONATOR

4.2.1   Short-Circuited $\lambda_g/2$ Transmission Line

Now, let us consider a finite-length lossy transmission line section with a short circuit at the right end, as shown in Figure 4.3a, where Z_0, β, and α are the characteristic impedance, propagation constant, and attenuation constant of the line, respectively. Such a transmission line may behave as a series resonant circuit when its length (l) is $\lambda_g/2$, or multiples of $\lambda_g/2$ at the frequency $\omega = \omega_0$. Figure 4.3b depicts its equivalent lumped-element resonant circuit. From (2.14), the input impedance of such a short-circuited transmission line can be written as

$$Z_{in} = Z_0 \tanh(\alpha + j\beta)l$$
$$= Z_0 \frac{\tanh \alpha l + j \tan \beta l}{1 + j \tan \beta l \tanh \alpha l}. \tag{4.1}$$

For series-type resonant circuit with lumped-elements shown in Figure 4.3b, we have

$$Z_{in} = R + j\omega L + \frac{1}{j\omega C}. \tag{4.2}$$

As we discussed in Chapter 3, the transmission line and its lumped-element resonant circuits exhibit the same resonance behavior. Plus, their slope parameters must be exactly the same at resonant frequency. From Equation (3.48), the reactance slope parameter x of the series-type resonator can be derived as

$$x = \frac{\omega_0}{2} \cdot \frac{dX}{d\omega}\bigg|_{\omega=\omega_0}, \tag{4.3}$$

where X is the reactance of the input impedance of the resonant circuit.

Assuming a TEM line has a small loss and $\alpha l \ll 1$, we have $\tanh \alpha l \approx \alpha l$ and $\tan \beta l \tanh \alpha l \ll 1$. Hence, Equation (4.1) becomes

$$Z_{in} = Z_0(\alpha l + j \tan \beta l). \tag{4.4}$$

Substituting Equations (4.2) and (4.4) into Equation (4.3) gives

$$
\begin{aligned}
x &= \frac{\omega_0}{2} \cdot \frac{d(X)}{d\omega}\bigg|_{\omega=\omega_0} \\
&= \frac{\omega_0}{2} \cdot \frac{d}{d\omega}\left(\omega L - \frac{1}{\omega C}\right)\bigg|_{\omega=\omega_0} = \frac{\omega_0}{2} \cdot \left(L - \frac{1}{\omega^2 C}\right)\bigg|_{\omega=\omega_0}, \\
&= \frac{\omega_0}{2} \cdot (L + L) = \omega_0 L
\end{aligned}
\tag{4.5a}
$$

and

$$
\begin{aligned}
x &= \frac{\omega_0}{2} \cdot \frac{d(X)}{d\omega}\bigg|_{\omega=\omega_0} = \frac{\omega_0}{2} \cdot \frac{l}{v_p} \cdot \frac{d(X)}{d\beta l}\bigg|_{l=n\lambda_g/2} \\
&= Z_0 \frac{\omega_0}{2} \cdot \frac{l}{v_p} \cdot \frac{d}{d\beta l}(\tan \beta l)\bigg|_{l=n\lambda_g/2} = Z_0 \frac{\omega_0}{2} \cdot \frac{l}{v_p} \cdot \sec^2 \beta l \bigg|_{l=n\lambda_g/2}. \\
&= Z_0 \frac{\omega_0}{2} \cdot \frac{n\lambda_g}{2v_p} \cdot \sec^2 \pi = \omega_0 \frac{nZ_0\pi}{2\omega_0}
\end{aligned}
\tag{4.5b}
$$

From Equations (4.2), (4.4), and (4.5), the elements of the lumped resonant circuit in Figure 4.3b can be expressed as a function of those in the initial transmission line circuit in Figure 4.3a:

$$R = Z_0 \alpha l = \frac{n}{2} Z_0 \alpha \lambda_g \tag{4.6a}$$

$$L = \frac{nZ_0\pi}{2\omega_0} \tag{4.6b}$$

$$C = \frac{1}{\omega_0^2 L} = \frac{2}{nZ_0\pi\omega_0}, \tag{4.6c}$$

where $n = 1, 2, 3 \ldots, l = n\lambda_g/2$ and $\omega_0^2 = 1/LC$.

Then, the Q-factor of this resonator can be defined as

$$Q = \frac{\omega_0 L}{R} = \frac{\pi}{\alpha \lambda_g} = \frac{\beta}{2\alpha}, \tag{4.7}$$

where $\beta l = \pi$ indicates the first resonance. It is of interest to note that Q is proportional to the phase coefficient while inversely proportional to the attenuation coefficient of a lossy transmission line.

4.2.2   Open-Circuited $\lambda_g/2$ Transmission Line

A transmission line with an open-circuited end can behave as a parallel-type resonator when its length is $\lambda_g/2$, or multiples of $\lambda_g/2$ at the frequency $\omega = \omega_0$. The transmission line circuit and its equivalent lumped-element resonant circuit are shown in Figure 4.4a,b.

The input admittance of this open-circuited transmission line with length l is

$$
\begin{aligned}
Y_{in} &= Y_0 \tanh (\alpha + j\beta) l \\
&= Y_0 \frac{\tanh \alpha l + j \tan \beta l}{1 + j \tan \beta l \tanh \alpha l}.
\end{aligned}
\tag{4.8}
$$

For its parallel type of lumped-element resonant circuit, we have

$$
Y_{in} = G + j\omega C + \frac{1}{j\omega L}.
\tag{4.9}
$$

From Equation (3.49), equivalence between these two resonant circuits can be established under the identical susceptance slope parameter b:

$$
b = \frac{\omega_0}{2} \cdot \frac{dB}{d\omega}\bigg|_{\omega=\omega_0}.
\tag{4.10}
$$

where B is the susceptance of the input admittance of the resonant circuit.

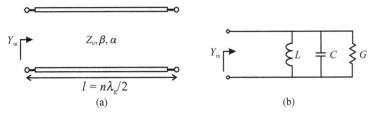

Figure 4.4. Resonant circuit models for an open-circuit lossy transmission line resonator with length l. (a) Transmission line resonant circuit. (b) Its equivalent parallel type lumped-element resonant circuit.

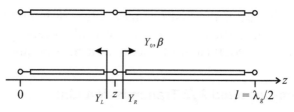

Figure 4.5. Resonant circuit model for an open-circuited lossless transmission line resonator with length l.

Since Equations (4.8)–(4.10) for a parallel GLC resonant circuit has the same fashion as Equations (4.1)–(4.3) for a series RLC resonant circuit, we just need to replace Z_0 with Y_0, R with G, and swap over L and C in Equation (4.5), so that

$$b = \frac{nY_0\pi}{2} \tag{4.11a}$$

$$G = Y_0\alpha l = \frac{n}{2}Y_0\alpha\lambda_g \tag{4.11b}$$

$$C = \frac{nY_0\pi}{2\omega_0} \tag{4.11c}$$

$$L = \frac{1}{\omega_0^2 C} = \frac{2}{nY_0\pi\omega_0}, \tag{4.11d}$$

where $n = 1, 2, 3 \ldots$ and then

$$Q = \frac{\omega_0 C}{G} = \frac{\pi}{\alpha\lambda_g} = \frac{\beta}{2\alpha} \tag{4.12}$$

EXAMPLE 4.1   Resonant condition of an open-circuited line resonator.

Consider a resonator shown in Figure 4.5 that consists of a $\lambda_g/2$ lossless transmission line terminated by open circuits at both ends. At an arbitrary point z along the line, derive the admittances $Y_L(z)$ and $Y_R(z)$ looking to the left and right ends, respectively. Show that the resonance condition of this line resonator must be $Y_L(z) + Y_R(z) = 0$.

Solution

From Equation (2.15),

$$Y_L(z) = jY_0 \tan \beta z$$
$$Y_R(z) = jY_0 \tan \beta(l - z).$$

At the first resonant frequency $\omega = \omega_0$, $\beta l = \pi$ is obtained.
Thus,

$$
\begin{aligned}
Y_L(z) + Y_R(z) &= jY_0 \tan \beta z + jY_0 \tan \beta(l - z) \\
&= jY_0 \tan \beta z + jY_0 \tan(\pi - \beta z) \\
&= jY_0 \tan \beta z - jY_0 \tan \beta z \\
&= 0.
\end{aligned}
$$

This condition is truly valid for any lossless resonator operating at resonance, and it is the foundation of the well-known transverse resonance technique [2].

4.2.3 Short-Circuited $\lambda_g/4$ Transmission Line

A short-circuited transmission line with length of $\lambda_g/4$, or odd multiples of $\lambda_g/4$, may also behave as a parallel resonant circuit, as shown in Figure 4.6. From Equation (4.1), the input admittance of this short-circuited transmission line with length l can be written as

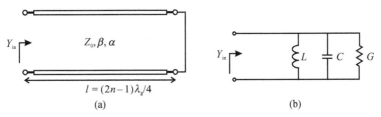

Figure 4.6. A short-circuited lossy transmission line resonator with the length of $(2n - 1)\lambda_g/4$ ($n = 1, 2, 3 \ldots$). (a) Transmission line resonant circuit. (b) Its equivalent parallel type of lumped-element resonant circuit.

$$Y_{in} = Y_0 \coth(\alpha + j\beta)l$$

$$= Y_0 \frac{\tanh \alpha l - j \cot \beta l}{1 - j \cot \beta l \tanh \alpha l} \qquad (4.13)$$

$$\approx Y_0(\alpha l - j \cot \beta l).$$

where we assume $\alpha l \ll 1$, and so $\tanh \alpha l \approx \alpha l$ and $\cot \beta l \tanh \alpha l \ll 1$. Substituting Equations (4.9) and (4.13) into Equation (4.10) gives

$$x = \frac{\omega_0}{2} \cdot \frac{d(X)}{d\omega}\bigg|_{\omega=\omega_0}$$

$$= \frac{\omega_0}{2} \cdot \frac{d}{d\omega}\left(\omega C - \frac{1}{\omega L}\right)\bigg|_{\omega=\omega_0} = \frac{\omega_0}{2} \cdot \left(C - \frac{1}{\omega^2 L}\right)\bigg|_{\omega=\omega_0} \qquad (4.14a)$$

$$= \frac{\omega_0}{2} \cdot (C + C) = \omega_0 C,$$

and

$$b = \frac{\omega_0}{2} \cdot \frac{d(B)}{d\omega}\bigg|_{\omega=\omega_0}$$

$$= -Y_0 \frac{\omega_0}{2} \cdot \frac{l}{v_p} \cdot \frac{d}{d\beta l}(\cot \beta l)\bigg|_{l=(2n-1)\lambda_g/4} = -Y_0 \frac{\omega_0}{2} \cdot \frac{l}{v_p} \cdot -\csc^2 \beta l\bigg|_{l=(2n-1)\lambda_g/4}$$

$$= Y_0 \frac{\omega_0}{2} \cdot \frac{(2n-1)\lambda_g}{2v_p} \cdot \csc^2 \frac{\pi}{2} = \omega_0 \frac{(2n-1)Y_0\pi}{4\omega_0}.$$

$$(4.14b)$$

From Equations (4.9), (4.13), and (4.14), equivalence between these two circuits leads to

$$G = Y_0 \alpha l = \frac{(2n-1)}{4} Y_0 \alpha \lambda_g \qquad (4.15a)$$

$$C = \frac{(2n-1)Y_0\pi}{4\omega_0} \qquad (4.15b)$$

$$L = \frac{1}{\omega_0^2 C} = \frac{4}{(2n-1)Y_0\pi\omega_0}, \qquad (4.15c)$$

where $n = 1, 2, 3 \ldots$ and then

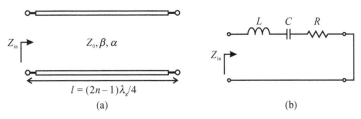

Figure 4.7. A open-circuited lossy transmission line resonator with the length of $(2n - 1)\lambda_g/4$ ($n = 1, 2, 3 \ldots$). (a) Transmission line resonant circuit. (b) Its equivalent series type of lumped-element resonant circuit.

$$Q = \frac{\omega_0 C}{G} = \frac{\pi}{\alpha \lambda_g} = \frac{\beta}{2\alpha}, \tag{4.16}$$

where $\beta l = \pi/2$ at the first resonance.

4.2.4   Open-Circuited $\lambda_g/4$ Transmission Line

The functionality of a series type of resonant circuit can also be achieved using an open-circuited transmission line of length $\lambda_g/4$, or odd multiples of $\lambda_g/4$, as shown in Figure 4.7. Similar to Equation (4.13), the input impedance of this open-terminated transmission line with length l becomes

$$\begin{aligned} Z_{in} &= Z_0 \coth(\alpha + j\beta)l \\ &= Z_0 \frac{\tanh \alpha l - j \cot \beta l}{1 - j \cot \beta l \tanh \alpha l} \\ &\approx Z_0(\alpha l - j \cot \beta l). \end{aligned} \tag{4.17}$$

Equivalence between these two circuits can be achieved by replacing Y_0 with Z_0, G with R, and swapping over L and C, such that

$$R = Z_0 \alpha l = \frac{(2n-1)}{4} Z_0 \alpha \lambda_g \tag{4.18a}$$

$$L = \frac{(2n-1)Z_0 \pi}{4\omega_0} \tag{4.18b}$$

$$C = \frac{1}{\omega_0^2 L} = \frac{4}{(2n-1)Z_0 \pi \omega_0}, \tag{4.18c}$$

where $n = 1, 2, 3 \ldots$ and then

$$Q = \frac{\omega_0 L}{R} = \frac{\pi}{\alpha \lambda_g} = \frac{\beta}{2\alpha}. \tag{4.19}$$

From the above analysis, we note that all the four types of resonators have the same unloaded Q-factor since the unloaded Q-factor of a resonator is only dependent on per-unit-length parameters of a transmission line regardless of its varied electrical length and terminations.

4.3 STEPPED IMPEDANCE RESONATORS

Figure 4.8 shows a half-wavelength open-circuited transmission line resonator with uniform and nonuniform width or impedance, that is, $Z_1 = Z_2$ or $Z_1 \neq Z_2$. For such a half-wavelength resonator with uniform width, its resonances will occur when

$$\theta_T = \theta_2 + 2\theta_1 + \theta_2 = n\pi, \tag{4.20}$$

where θ_T is the total electrical length of the resonator and n is the number of resonant modes.

From Equation (4.20), we can understand that the second resonance of $n = 2$ always occurs at twice of the first resonant frequency, and it is referred to as the first harmonic frequency response with reference to

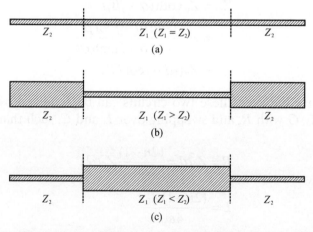

Figure 4.8. Configurations of a half-wavelength open-circuited transmission line reso-nators with different impedance ratio $R_Z = Z_1/Z_2$. (a) UIR ($R_Z = 1$). (b) SIR, type I ($R_Z < 1$). (c) SIR, type II ($R_Z > 1$).

its fundamental one. As one of main drawbacks, the UIR has several undesired spurious resonant modes, occurring at integer multiples of its first resonant frequency, which could be harmful for the design of single-band bandpass filter. To solve this problem, one of the most popular methods is to increase the ratio of the lowest higher-order resonant frequencies to its first counterpart by forming a nonuniform impedance resonator or SIR [3], as shown in Figure 4.8b,c. The further discussion on SIR will be addressed later on for designing a variety of filters.

4.3.1 Fundamental Properties of SIR

Figure 4.9 shows some typical structures of the open-circuited SIRs with two low-impedance sections at two ends and a high-impedance section at the center. By splitting or bending the two low-impedance sections of a SIR, the split-end and hair-pin types of SIR can be obtained from its original straight type of geometry, as shown in Figure 4.9b,c. For the structures shown in Figure 4.9d–f, additional internal coupling between the edges of the low-impedance lines has to be involved. In the literature, these resonators have been widely used in microwave

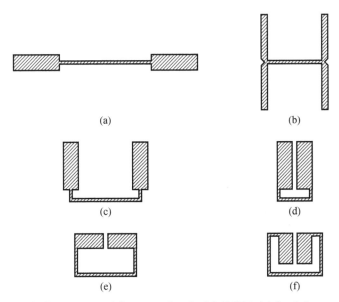

Figure 4.9. Typical structures of the open-circuited $\lambda_g/2$ SIR. (a) Straight type. (b) Split-end type. (c) U-bended or hairpin type. (d) Hairpin type with internal coupling. (e) Ring type. (f) Ring type with internal coupling.

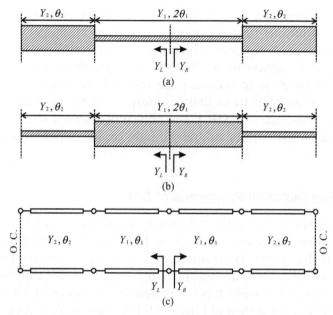

Figure 4.10. Configurations of a SIR and its equivalent transmission line circuit. (a) Type I with $R_Z < 1$. (b) Type II with $R_Z > 1$. (c) Equivalent transmission line circuit.

circuit design, especially for the design of microwave bandpass filters due to its unique resonant characteristic.

As mentioned in Section 4.2.2, an open-circuited transmission line with a length of $n\lambda_g/2$ will behave as a parallel-type resonator. Therefore, it is convenient to use the characteristic admittances instead of the characteristic impedances in Figure 4.8 for our analysis herein. Figure 4.10 shows two basic structures of SIR (type I and type II) and their equivalent transmission line circuits. Based on the resonant condition discussed in Example 4.1, the input admittances, Y_L and Y_R, looking to the left and right sides at the center of the SIR, must satisfy

$$Y_L + Y_R = 0. \tag{4.21}$$

From Example 2.2, we have

$$
\begin{aligned}
Y_L &= jY_1 \frac{Y_1 \tan\theta_1 + Y_2 \tan\theta_2}{Y_1 - Y_2 \tan\theta_1 \tan\theta_2} \\
&= jY_1 \frac{R_Z \tan\theta_1 + \tan\theta_2}{R_Z - \tan\theta_1 \tan\theta_2}
\end{aligned}
\tag{4.22}
$$

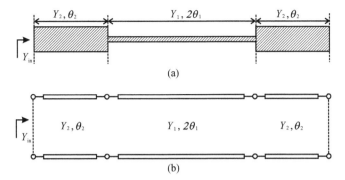

Figure 4.11. (a) Topology of a SIR. (b) Its equivalent transmission line circuit.

where R_Z is the ratio of two characteristic impedances, that is, $R_Z = Y_1/Y_2 = Z_2/Z_1$.

Since the structure is symmetrical, $Y_L = Y_R$. Substituting Equation (4.21) into Equation (4.22) and multiplying the denominator at both sides gives

$$2jY_1(R_Z - \tan\theta_1 \tan\theta_2)(R_Z \tan\theta_1 + \tan\theta_2) = 0. \tag{4.23}$$

As proved in Example 4.1, the resonant condition is still valid if the standing position is selected at one end of the resonator. As shown in Figure 4.11, the input admittance (Y_{in}) at the left end, looking into the right side, can be derived as

$$Y_{in} = jY_2 \frac{2(R_Z \tan\theta_1 + \tan\theta_2)(R_Z - \tan\theta_1 \tan\theta_2)}{R_Z(1 - \tan^2\theta_1)(1 - \tan^2\theta_2) - 2(1 + R_Z^2)\tan\theta_1 \tan\theta_2}. \tag{4.24}$$

In this context, the resonant condition at resonances leads to

$$Y_{in} = 0. \tag{4.25}$$

As a result, even- and odd-mode resonant frequencies can be solved via (4.23) or (4.25) as

$$R_Z - \tan\theta_1 \tan\theta_2 = 0 \quad \text{(at odd-mode frequencies)} \tag{4.26a}$$

$$R_Z \tan\theta_1 + \tan\theta_2 = 0 \quad \text{(at even-mode frequencies)}. \tag{4.26b}$$

At the first resonance, the total electrical length of the SIR is derived as

$$\theta_T = 2\theta_1 + 2\tan^{-1}\left(\frac{R_Z}{\tan\theta_1}\right). \tag{4.27}$$

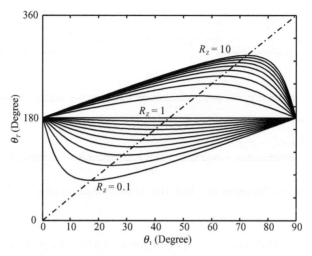

Figure 4.12. Resonant condition of the SIR at its first resonance.

The above analysis shows that a smaller impedance ratio R_z results to a smaller total electrical length of the resonator, so the resonator length may be reduced to a certain extent. Figure 4.12 shows the relationship between the electrical length θ_1 and total resonator length θ_T as the impedance ratio R_Z is changed from 0.1 to 10. We find that the total electrical length θ_T reaches to a maximum value when $R_Z > 1$ and a minimum value when $R_Z < 1$. These extreme values can be determined by substituting $\theta_2 = \theta_T/2 - \theta_1$ back into Equation (4.26b) as

$$\tan\frac{\theta_T}{2} = \frac{R_Z + \tan^2\theta_1}{(1 - R_z)\tan\theta_1}$$

$$= \frac{\sqrt{R_Z}}{1 - R_z}\cdot\left(\frac{\tan\theta_1}{\sqrt{R_Z}} + \frac{\sqrt{R_Z}}{\tan\theta_1}\right). \tag{4.28}$$

Then, the maximum or minimum value of θ_T is given by

$$\theta_T^{\text{max|min}} = 2\tan^{-1}\frac{2\sqrt{R_Z}}{1 - R_z}, \tag{4.29}$$

where $\theta_1 = \theta_2 = \tan^{-1}\sqrt{R_Z}$.

Based on Equation (4.26), a set of resonant frequencies $(f_1, f_2, \ldots)$ can be determined from θ_1 and θ_2. For a special case where $\theta_1 = \theta_2 = \theta$, we have

$$\theta(f_1) = \tan^{-1}\sqrt{R_Z} \qquad\qquad (4.30a)$$

$$\theta(f_2) = \frac{\pi}{2} \qquad\qquad (4.30b)$$

$$\theta(f_3) = \pi - \tan^{-1}\sqrt{R_Z} \qquad\qquad (4.30c)$$

$$\theta(f_4) = \pi. \qquad\qquad (4.30d)$$

Thus,

$$\frac{f_2}{f_1} = \frac{\theta(f_2)}{\theta(f_1)} = \frac{\pi}{2\tan^{-1}\sqrt{R_Z}} \qquad\qquad (4.31a)$$

$$\frac{f_3}{f_1} = \frac{\theta(f_3)}{\theta(f_1)} = \frac{\pi}{\tan^{-1}\sqrt{R_Z}} - 1 \qquad\qquad (4.31b)$$

$$\frac{f_4}{f_1} = \frac{\theta(f_4)}{\theta(f_1)} = \frac{\pi}{\tan^{-1}\sqrt{R_Z}}. \qquad\qquad (4.31c)$$

Figure 4.13 plots the three graphs of normalized frequencies, that is, f_2/f_1, f_3/f_1, and f_4/f_1, versus R_Z. With reference to f_1, both higher-order resonances depart from the first resonance as R_Z decreases from unity ($10^0 = 1$). As a result, the upper stopband of a single-band SIR bandpass filter when $R_Z < 1$ can be widened. On the other hand, as R_Z increases

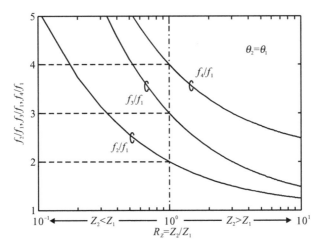

Figure 4.13. Normalized resonant frequencies versus impedance ratio of an SIR ($R_Z = Z_2/Z_1$).

TABLE 4.1   Electrical Length θ_T of SIR at Resonances

	$R_Z = 0.5$		$R_Z = 1$		$R_Z = 2$	
Resonance	θ	θ_T	θ	θ_T	θ	θ_T
1st (f_1)	35.26°	141.06°	45°	180°	54.74°	218.94°
2nd (f_2)	90°	360°	90°	360°	90°	360°
3rd (f_3)	144.74°	578.94°	135°	540°	125.26°	501.06°
4th (f_4)	180°	720°	180°	720°	180°	720°

from unity, the first two higher-order resonances both move down in frequency toward the first resonance. This property of a SIR filter with $R_Z > 1$ was previously used to realize a low-loss diplexer on a very high dielectric substrate [4]. In References 5 and 6, application of the SIR with $R_Z > 1$ was explored toward a new type of wide-band bandpass filters that simultaneously utilize the first few resonances. Using this initial single resonator, four or five transmission poles were reportedly achieved in the desired passband. In the following, we will focus our attention on proposal, design, and implementation of these wideband filters based on multiple-mode resonator (MMR).

To simplify our analysis, three resonators with $R_Z = 0.5$, 1, and 2 under the assumption of $\theta_1 = \theta_2 = \theta$ are considered at first. The calculated total electrical length θ_T at different resonance is tabulated in Table 4.1. We can find that the total electrical length ($\theta_T = 4\theta$) at the second resonant frequency of f_2 is always fixed at 2π or 360°, representing a one-wavelength resonance at this frequency. In particular, θ_T at the first resonant frequency is in direct proportion with R_Z, and it always has a minimum value when $R_Z < 1$ due to (4.29).

By setting the first resonant frequency at 1.0 GHz and forming a capacitive weak coupling structure at two ends of each resonator, the frequency responses of $|S_{21}|$ can be calculated, and they are plotted in Figure 4.14. It can be seen that the higher-order resonant frequencies of this constituted SIR varies as a function of the impedance ratio of R_Z. This unique feature is very useful in the design of a variety of bandpass filters. On the one hand, the higher-order harmonic frequencies of SIR can be pushed far away from the fundamental frequency so as to make up a single-band SIR bandpass filter with good upper-stopband performance [3]. On the other hand, they can properly be reallocated so as to build up a new class for multiband [7, 8] or wide-band bandpass filters [9], as will be discussed in the later sections.

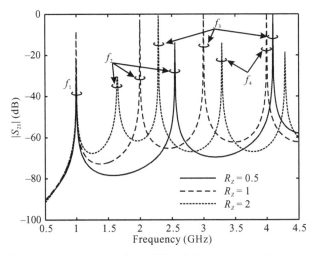

Figure 4.14. Frequency responses of the SIR with versus impedance ratio of an SIR $(R_Z = Z_2/Z_1)$.

4.3.2 Slope Parameters of SIR

In practical synthesis design of a filter (referring to Section 3.4), it usually starts with the transformation of a distributed-element resonator to its equivalent lumped-element RLC or GLC resonator. At the first resonance, the SIR can be equivalently expressed as a parallel-type lumped-element resonator, where each lumped element can be derived from the slope parameters of a resonator, as what we have done in Section 4.2.2. In the case of $\theta_2 = \theta_1 = \theta$, Equation (4.24) is simplified as

$$Y_{in} = jY_2 \frac{2\tan\theta(1+R_Z)(R_Z - \tan^2\theta)}{R_Z - 2(1+R_Z + R_Z^2)\tan^2\theta + R_Z \tan^4\theta}. \qquad (4.32)$$

From Equation (4.14), the susceptance slope parameter of SIR can be derived as

$$\begin{aligned}
b &= \frac{\omega_0}{2} \cdot \frac{dB}{d\omega}\bigg|_{\omega=\omega_0} \\
&= \frac{\theta_0}{2} \cdot \frac{d\,\mathrm{Im}(Y_{in})}{d\theta}\bigg|_{\theta=\theta_0} \\
&= \frac{\theta_0}{2} \cdot 2Y_2(1+R_Z) \cdot \frac{2}{1+R_Z} \\
&= 2\theta_0 Y_2.
\end{aligned} \qquad (4.33)$$

For the UIR shown in Figure 4.8a, $\theta_0 = \pi/4$ and $Y_1 = Y_2 = Y_0$, we can obtain an equation that is the exactly same as Equation (4.11a) when $n = 1$:

$$b = \frac{Y_0\pi}{2} \tag{4.34}$$

From Equation (4.5a), each element of the parallel-type or GLC resonator can then be determined with virtue of its susceptance slope parameter (b) as

$$L = \frac{1}{\omega_0 b} = \frac{1}{2\omega_0\theta_0 Y_2} \tag{4.35a}$$

$$C = \frac{b}{\omega_0} = \frac{2\theta_0 Y_2}{\omega_0} \tag{4.35b}$$

$$G = \frac{b}{Q} = \frac{2\theta_0 Y_2}{Q} \tag{4.35c}$$

where Q is the unloaded-Q of the SIR, and it is dependent on not only the substrate but also the stepped impedance ratio of the constituted SIR, as will be discussed later on.

4.3.3 Design of the Parallel-Coupled Line Bandpass Filters with SIR

The basic configuration of a parallel-coupled line bandpass filter is composed of single- or multiple-stage SIRs are illustrated in Figure 4.15. For synthesis design of a single-band bandpass filter, the traditional design procedure in Section 3.5 can be followed. First, we assume that the susceptance slope parameters of each stage of the parallel-coupled lines have the same value of $b = 2\theta_0 Y_0$. For the known element value g_j and specified fractional bandwidth (FBW), the admittance inverter parameter $J_{k,k+1}$ can be obtained from Figure 3.12b as

$$J_{01} = \sqrt{\frac{Y_0 b_1 FBW}{g_0 g_1}} = Y_0\sqrt{\frac{2FBW\theta_0}{g_0 g_1}} \tag{4.36a}$$

$$J_{k,k+1}\big|_{k=1\,\text{to}\,n-1} = \sqrt{\frac{b_k b_{k+1}}{g_k g_{k+1}}} = Y_0\frac{2FBW\theta_0}{\sqrt{g_k g_{k+1}}} \tag{4.36b}$$

$$J_{n,n+1}\big|_{k=1\,\text{to}\,n-1} = \sqrt{\frac{Y_0 b_1 FBW}{g_n g_{n+1}}} = Y_0\sqrt{\frac{2FBW\theta_0}{g_n g_{n+1}}}. \tag{4.36c}$$

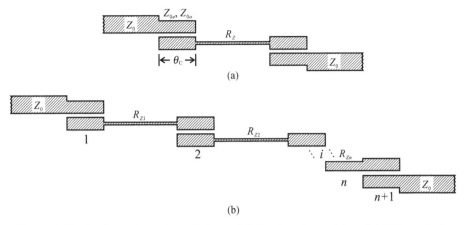

Figure 4.15. Configuration of parallel-coupled line bandpass filters. (a) With a single SIR. (b) With multiple SIRs.

EXAMPLE 4.2 Design a bandpass filter with the following parameters and specifications:

Fundamental frequency f_0	2.0 GHz
Number of resonators	3
Response	Chebyshev
Ripple level L_{Ar}	0.1 dB
FBW	5%
Impedance ration R_Z	0.5
Z_2	50

Solution

From Table 3.2, the element values of a Chebyshev lowpass prototype with $L_{Ar} = 0.1$ dB and $n = 5$ can be found as

$$g_0 = g_6 = 1 \quad g_1 = g_5 = 1.1468$$
$$g_2 = g_4 = 1.3712 \quad g_3 = 1.9750.$$

From Equation (4.30a), we have $\theta_0 = 35.26°$. Then, the normalized admittance inverter parameters, even- and odd-mode impedances are calculated by Equations (4.36) and (3.57), and their values are tabulated in the following table:

n	$J_{n,n+1}/Y_0$	Z_{0e} (Ω)	Z_{0o} (Ω)
0 and 5	0.2317	81.50	36.54
1 and 4	0.0491	54.63	46.09
2 and 3	0.0374	53.46	46.96

Once the impedances are known, the line width and gap size for the selected substrate can be found using the closed-form design formula [10]. At the center frequency of 2.0 GHz, electrical length of all the three parallel-coupled sections is set as θ_0. Until now, such a three-stage SIR filter is completely designed. Figure 4.16 plots its frequency responses. We can find that the second harmonic frequency (f_2) has been shifted up from 4.0 to 5.1 GHz, thus widening the upper stopband bandwidth.

4.4   MULTIPLE-MODE RESONATOR

4.4.1   Open-Circuited Multiple-Mode Resonator

As illustrated in Figure 4.13, the higher-order resonances can be moved toward the first-order resonance of a SIR as R_Z increases in the case of $R_Z > 1$. These multiple resonances could be simultaneously excited and utilized together to make up a wide passband in a frequency range covering these multiple resonant frequencies, so that the above resona-

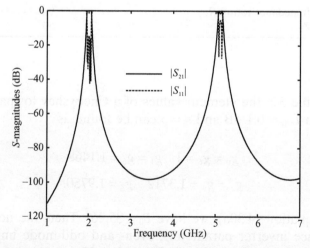

Figure 4.16. Simulated frequency responses of the designed parallel-coupled line SIR bandpass filter.

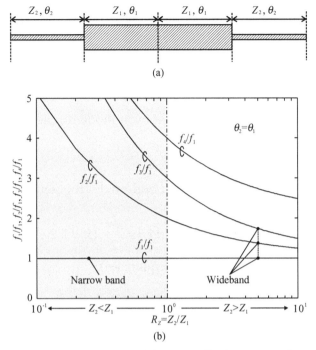

Figure 4.17. Geometry of an SIR and its normalized resonant frequencies when $\theta_2 = \theta_1$.

tor with similar configuration as that in Figure 4.13 is named as multiple-mode resonator (MMR) [9]. In our early works [5, 6], the first two resonant modes were excited and utilized to form a passband with about 70% *FBW*. Unlike the traditional single-mode resonator filters, the MMR-based filters are constructed by allocating the first few resonant modes into the desired wide passband and quasi-equally exciting their amplitudes via quarter-wavelength coupled lines at center frequency.

Figure 4.17 shows the geometry of an open-circuited SIR when $R_Z > 1$ and the normalized resonant frequencies for the case of $\theta_2 = \theta_1$. The resonant frequencies of this SIR can be determined by Equation (4.26), which is written here when $\theta_2 = \theta_1 = \theta$:

$$\theta = \tan^{-1}\sqrt{R_Z} \qquad \text{(at odd-mode frequencies)} \qquad (4.37a)$$

$$\theta = \frac{k\pi}{2}, (k = 1, 2, \ldots) \quad \text{(at even-mode frequencies)}. \qquad (4.37b)$$

From Figure 4.17, it can be seen that the first three resonances are moved closely to each other as R_Z increases, and they can be utilized to construct a wide passband. Meanwhile, the fourth resonance brings out the first spurious frequency response above the core wide passband.

For the case of $\theta_2 = 2\theta_1 = 2\theta$, Equation (4.26) can be expressed as

$$R_Z - \tan\theta_1 \tan 2\theta_1 = 0 \quad \text{(at odd-mode frequencies)} \quad (4.38a)$$

$$R_Z \tan\theta_1 + \tan 2\theta_1 = 0 \quad \text{(at even-mode frequencies)}, \quad (4.38b)$$

or

$$\theta = \tan^{-1}\sqrt{\frac{R_Z}{2+R_Z}} \quad \text{(at odd-mode frequencies)} \quad (4.39a)$$

$$\theta = \tan^{-1}\sqrt{\frac{R_Z+2}{R_Z}}, \text{ and } \theta = \frac{k\pi}{2}, (k = 1, 2, \ldots)$$
$$(4.39b)$$
$$\text{(at even-mode frequencies)}.$$

So, at the first three resonances, we have

$$\theta(f_1) = \tan^{-1}\sqrt{\frac{R_Z}{R_Z+2}} \quad (4.40a)$$

$$\theta(f_2) = \tan^{-1}\sqrt{\frac{R_Z+2}{R_Z}} \quad (4.40b)$$

$$\theta(f_3) = \frac{\pi}{2}. \quad (4.40c)$$

Therefore, the normalized resonant frequencies are given as

$$\frac{f_2}{f_1} = \frac{\theta(f_2)}{\theta(f_1)} = \frac{\tan^{-1}\sqrt{\dfrac{R_Z+2}{R_Z}}}{\tan^{-1}\sqrt{\dfrac{R_Z}{R_Z+2}}} \quad (4.41a)$$

$$\frac{f_3}{f_1} = \frac{\theta(f_3)}{\theta(f_1)} = \frac{\pi}{2\tan^{-1}\sqrt{\dfrac{R_Z}{R_Z+2}}}. \quad (4.41b)$$

Figure 4.18 plots the first three normalized resonant frequencies as a function of the impedance ratio R_Z. In this case, the second resonant frequency drops off towards the first one much faster than the third one. As a result, the first two resonances could occur in proximity to

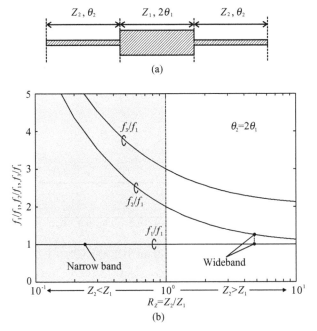

Figure 4.18. Geometry of an SIR and its normalized resonant frequencies when $\theta_2 = 2\theta_1$.

each other as R_Z increases, and they are both excited to be able to launch a wide passband as done in References 5 and 6.

To give a close investigation on the controllable multiple-resonance behavior of the MMR, an SIR with arbitrary lengths of low- and high-impedance line sections is considered as below. For a general purpose, a length ratio of two distinct sections is defined as

$$u = \frac{\theta_2}{\theta_1 + \theta_2}. \tag{4.42}$$

Then, Equation (4.37) can be expressed as

$$R_Z - \tan\theta_1 \tan\frac{u}{1-u}\theta_1 = 0 \quad \text{(at odd-mode frequencies)} \tag{4.43a}$$

$$R_Z \tan\theta_1 + \tan\frac{u}{1-u}\theta_1 = 0 \quad \text{(at even-mode frequencies).} \tag{4.43b}$$

When $R_Z = 5$, the normalized resonant frequencies can be obtained as a function of the length ratio u, and they are plotted in Figure 4.19.

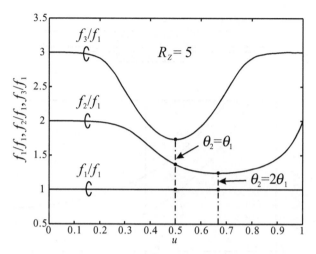

Figure 4.19. Resonant spectrum of an SIR when $R_Z > 1$, or an MMR.

When u equals to 0 or 1, the SIR represents a UIR with Z_1 or Z_2. As u is increased from 0 to 1, both the second and third resonant frequencies shift down and move towards the first resonance, and they are then raised beyond a certain u toward the starting frequencies in the uniform case. Looking closely at Figure 4.19, we can find out that if $\theta_2 = \theta_1$ and $\theta_2 = 2\theta_1$ are selected, the first three or two resonant frequencies can be reasonably reallocated closely with each other. It implies that the MMR itself can bring out three or two transmission poles in these two cases.

4.4.2 Short-Circuited Multiple-Mode Resonator

Figure 4.20 shows another MMR with two short-circuited ends. At resonances, the input impedances, Z_L and Z_R, looking into the left- and right-hand sides from the center of the MMR, must satisfy

$$Z_L + Z_R = 0. \tag{4.44}$$

Then, we have

$$
\begin{aligned}
Z_L &= jZ_1 \frac{Z_1 \tan\theta_1 + Z_2 \tan\theta_2}{Z_1 - Z_2 \tan\theta_1 \tan\theta_2} \\
&= jZ_1 \frac{R_Y \tan\theta_1 + \tan\theta_2}{R_Y - \tan\theta_1 \tan\theta_2}.
\end{aligned}
\tag{4.45}
$$

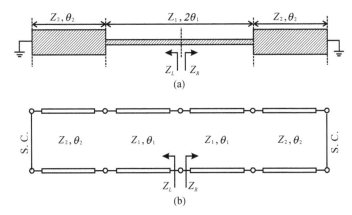

Figure 4.20. An MMR with short-circuited ends and its equivalent transmission line circuit.

where $R_Y = 1/R_Z = Z_1/Z_2$. Because $Z_L = Z_R$, we have

$$jZ_1(R_Y - \tan\theta_1 \tan\theta_2)(R_Y \tan\theta_1 + \tan\theta_2) = 0. \qquad (4.46)$$

The resonant frequencies can be determined by Equation (4.46) as

$$R_Y - \tan\theta_1 \tan\theta_2 = 0 \quad \text{(at odd-mode frequencies)} \qquad (4.47a)$$

$$R_Y \tan\theta_1 + \tan\theta_2 = 0 \quad \text{(at even-mode frequencies)}. \qquad (4.47b)$$

Comparing Equations (4.26) and (4.47), the only difference in resonant condition between the short- and open-circuited SIRs is the inverse relationship of impedance ratio, that is, swapping over impedance ratio and admittance ratio of two distinct sections of the MMRs. As such, $R_Z < 1$ and $R_Z > 1$ need to be separately selected for making up the short-circuited and open-circuit MMRs. With no repeated analysis, a few sets of design graphs shown in Figure 4.18b, Figure 4.19b, and Figure 4.20 for the open-circuited MMR can be straightforwardly applied for the analysis and design of the short-circuited MMR if R_Z is inversed or replaced by $1/R_Z$.

4.4.3 Design of the Parallel-Coupled Line Bandpass Filters with MMRs

The basic configuration of a parallel-coupled line bandpass filter using open- or short-circuited MMRs are shown in Figure 4.21. For a filter

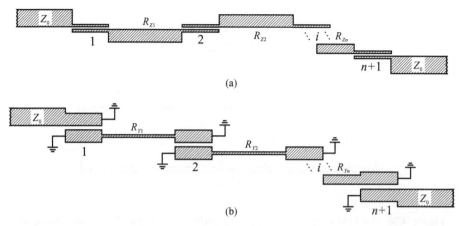

Figure 4.21. Configuration of parallel-coupled line bandpass filters using SIRs. (a) Open-circuited ends. (b) Short-circuited ends.

with n^{th}-order Chebyshev frequency response, the insertion loss function can be expressed from Equations (2.33) and (3.2a) as

$$L_A = 10\log_{10}|S_{21}|^{-2} = 10\log_{10}\left[1 + \varepsilon^2 T_n^2(x)\right]. \tag{4.48}$$

where ε^2 specifies the in-band ripple level and $T_n(x)$ is the nth-order first kind Chebyshev polynomial that oscillates between -1 and 1. For the kth pole of Equation (4.48), the zero of $T_n(x)$ is given by Reference 11:

$$x_k = \cos\left(\frac{2n+1-2k}{2n}\pi\right), \quad (k = 1, 2, \ldots, n). \tag{4.49}$$

For an nth-order equal ripple-level Chebyshev bandpass filter with a specified FBW, the pole frequency f_k can be easily obtained by

$$\frac{f_k}{f_0} = 1 + x_k \times \frac{FBW}{2}, \tag{4.50}$$

where f_0 is the center frequency of the passband.

For the design of a parallel-coupled line bandpass filter using single-mode resonators, n resonators are essentially cascaded so as to match n transmission poles of Equation (4.50), and its value needs to be increased if a wideband filter is preferred. However, this procedure may degrade the filtering performance to a certain extent, such as higher insertion loss in the passband.

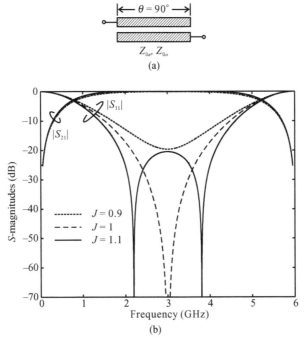

Figure 4.22. Parallel-coupled lines with different coupling effects.

As briefed early, multiple resonant modes in a single MMR could also be utilized to make up a wide passband instead of multiple numbers of single-mode resonators. Generally speaking, for a resonator using the first m resonant modes, only $(n - 2)/m$ resonators are needed to realize n transmission poles, where additional two transmission poles can be generated by tightly coupled lines at input and output ports [5, 6].

Consider a parallel-coupled line with two open-circuited ends shown in Figure 4.22a. When $\theta = 90°$, its image impedance can be simply obtained from Equations (2.75) and (3.55):

$$Z_i = \bar{J}Z_0, \tag{4.51}$$

where $\bar{J}$ is the normalized J-susceptance given by

$$\bar{J} = \frac{J}{Y_0} = \frac{Z_{0e} - Z_{0o}}{2Z_0}. \tag{4.52}$$

For a coupled line section with the input/output impedance of Z_0, Equation (4.51) reveals that good impedance matching for this two-port network is only dependent on $\bar{J}$ at center frequency, that is,

$$\begin{cases} \bar{J} < 1 \Rightarrow Z_i < Z_0 \text{ (mismatch)} \\ \bar{J} = 1 \Rightarrow Z_i = Z_0 \text{ (match)} \\ \bar{J} > 1 \Rightarrow Z_i > Z_0 \text{ (overmatch)} \end{cases} \qquad (4.53)$$

Figure 4.22 plots the S-magnitudes of a coupled-line section with one quarter wavelength at 3.0 GHz. When $\bar{J} = 1$, the coupled line is perfectly matched with one transmission pole at center frequency. When $\bar{J} = 0.9$, the coupled line is not well matched with ill-behaved transmission poles due to its weak coupling. When $\bar{J} = 1.1$, a single transmission pole is now split to two poles by enhancing the coupling degree of two coupled lines at the input and output ports. For a parallel-coupled bandpass filter with narrow or moderate bandwidth, $\bar{J}$ is usually not larger than unity. However, for a wideband bandpass filter using MMR, the overenhanced coupling is essentially required so as to create two additional transmission poles in the operating passband.

To realize a wideband microstrip filter in practice, the strip and slot widths of the parallel-coupled microstrip line (PCML) need to be reduced to achieve a tight coupling degree. However, this procedure may reduce the Q-factor due to the increased conductor loss. To circumvent this problem, an aperture is formed on the ground plane underneath the coupled strip conductor, as shown in Figure 4.23a, to offer an alternative PCML with enhanced coupling degree. Figure 4.23b exhibits that a single transmission pole can be split to the two poles as the aperture width is raised from $W = 2.2$ mm to $W = 3.0$ mm [5]. By combining this aperture-backed PCML and above-discussed multiple-mode resonators, a class of compact and wideband filters can be realized.

An initial design example of MMR-based filter is presented in References 5 and 6. Figure 4.24a shows the geometry of the constituted wideband bandpass filter where the coupled strip conductors in the two sides of a single multiple-mode resonator are backed by apertures. In this case, a two-mode resonator with $\theta_2 = 2\theta_1$ is employed, and a pair of open-circuited stubs are installed in shunt at the center. By taking advantages of loading stubs, the harmonic frequency (f_3) that appeared above twice of first resonant frequency can be suppressed. Figure 4.24b describes the simulated and measured frequency responses of the

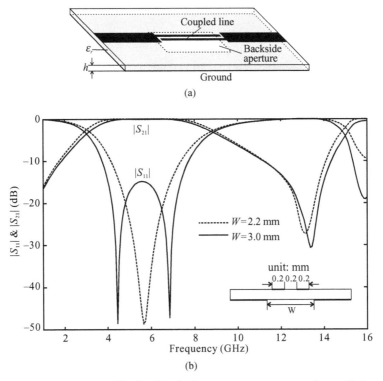

Figure 4.23. Geometry and simulated frequency responses of parallel-coupled microstrip line with a backside aperture. (a) Geometry. (b) $|S_{11}|$ and $|S_{21}|$ of PCML with different aperture widths.

designed four-pole bandpass filter with a wide *FBW* of about 60%. Both the simulated and measured insertion losses $|S_{21}|$ are found below 0.6 dB within the passband. Meanwhile, the simulated return loss $|S_{11}|$ is lower than 20 dB against the measured $|S_{11}|$ lower than 16 dB. The total length of this stub-loaded bandpass filter is about $0.7\lambda_g$ at 6.1 GHz, against $1.25\lambda_g$ for a conventional four-pole PCML bandpass filter as in Reference 1.

4.5 SUMMARY

In this chapter, we have investigated the UIR and SIR for a variety of bandpass filters. Equivalence of transmission line resonators and lumped-element *RLC* or *GLC* resonators are proved so as to describe a systematic synthesis design procedure for single-mode resonator

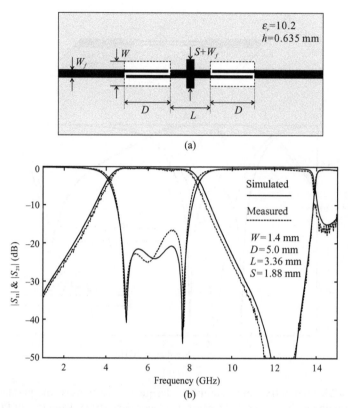

Figure 4.24. Geometry and simulated frequency responses of parallel-coupled microstrip line with a backside aperture. (a) Geometry. (b) $|S_{11}|$ and $|S_{21}|$ of PCML with different aperture widths.

based bandpass filters. After unique properties of SIR are discussed and demonstrated, the concept of MMR is introduced and applied to design a class of wideband filters using the first few resonant modes in a single MMR.

REFERENCES

1   G. Matthaei, L. Young, and E. M. T. Jones, *Microwave Filters, Impedance-Matching Network, and Coupled Structures*, Artech House, Dedham, MA, 1980.

2   D. M. Pozar, *Microwave Engineering*, 2nd ed., Section 3.9, John Wiley & Sons, Inc., New York, 1998.

3 M. Makimoto and S. Yamashita, "Bandpass filters using parallel coupled stripline stepped impedance resonators," *IEEE Trans. Microw. Theory Tech.* MTT-28(12) (1980) 1413–1417.

4 A. Sheta, J. P. Coupez, G. Tanne, S. Toutain, and J. P. Blot, "Miniature microstrip stepped impedance resonator bandpass filter and diplexers for mobile communications," *IEEE MTT-S Int. Dig.*, June 1996, Vol. 2, pp. 607–610.

5 L. Zhu, H. Bu, and K. Wu, "Aperture compensation technique for innovative design of ultra-broadband microstrip bandpass filter," *IEEE MTT-S Int. Dig.*, vol. 1 2000, pp. 315–318.

6 L. Zhu, H. Bu, and K. Wu, "Broadband and compact multi-mode microstrip bandpass filters using ground plane aperture technique," *IEE Proc. Microw. Antennas Propag* 149(1) (2002) 71–77.

7 S. Sun and L. Zhu, "Coupling dispersion of parallel-coupled microstrip lines for dual-band filters with controllable fractional pass bandwidths," *IEEE MTT-S Int. Dig.*, June 2005, pp. 2195–2198.

8 S. Sun and L. Zhu, "Compact dual-band microstrip bandpass filter without external feeds," *IEEE Microw. Wireless Compon. Lett.* 15(10) (2005) 644–646.

9 S. Sun and L. Zhu, "Multiple-mode-resonator-based bandpass filters," *IEEE Microwave Magazine* 10(2) (2009) 88–98.

10 B. C. Wadell, *Transmission Line Design Handbook*, Artech House, Norwood, MA, 1991.

11 Y.-C. Chiou, J.-T. Kuo, and E. Cheng, "Broadband quasi-Chebyshev bandpass filter with multimode stepped-impedance resonators (SIRs)," *IEEE Trans. Microw. Theory Tech.* 54 (2006) 3352–3358.

CHAPTER 5

MMR-BASED UWB BANDPASS FILTERS

5.1 INTRODUCTION

The traditional filter theory was systematically established under the assumption of narrow passband, and it has been found very powerful in the design of microwave filters with various filtering performance. For a wideband filter, it was usually constructed by cascading a few transmission line resonators via enhanced coupling. However, this traditional approach requires extremely high-degree coupling of coupled-lines between two adjacent resonators, resulting in a narrow stopband between the desired passband and its first harmonic passband. Since the FCC releases the ultra-wideband (UWB) spectrum for unlicensed commercial applications in 2002, tremendous attention has been paid on proposal and implementation of a variety of advanced filters with desired UWB specifications. But, unfortunately, a lot of critical issues still puzzle researchers in many aspects, such as good in-band transmission, highly rejected skirt, wide out-of-band attenuation, multiple-notched bands, and so on.

The concept of a multiple-mode resonator (MMR) with stepped impedance configuration was originally presented in Reference 1. As

Microwave Bandpass Filters for Wideband Communications, First Edition. Lei Zhu, Sheng Sun, Rui Li.
© 2012 John Wiley & Sons, Inc. Published 2012 by John Wiley & Sons, Inc.

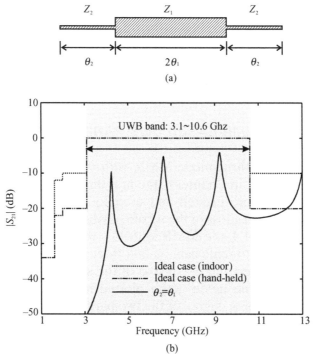

Figure 5.1. A MMR and its simulated $|S_{21}|$ within the desired FCC limits under weak coupling.

discussed in Chapter 4, the first two resonant modes were excited and utilized together to form a passband with about 70% fractional bandwidth (*FBW*). Later on, in 2005, this work was extended to make a UWB bandpass filter with a *FBW* of about 109.5% [2]. This initial UWB bandpass filter was built up using the microstrip line topology and can be easily migrated to be with other transmission lines. For a UWB band from 3.1 to 10.6 GHz, we can allocate the first three resonant frequencies into the UWB passband and set the coupling peak of a quarter-wave parallel-coupled line at center frequency, where two additional poles will be introduced. The simulated $|S_{21}|$ of a MMR circuit is shown in Figure 5.1. It is driven via capacitive weak coupling at two ends under the fixed $\theta_1 = \theta_2 = \pi/2$ at the center frequency 6.85 GHz, that is, the 2nd resonant frequency of the SIR in Section 4.3. The first three resonant frequencies are equally allocated within the FCC-specified UWB limits. It can be imagined that the desired five-pole ultra-wide passband can be formed as long as the coupling degree of two driven lines is properly increased. This chapter presents their varied geometries on microstrip lines, CPW, and hybrid structures. In order to further improve the filter performance, a class of MMR-based UWB filters is proposed and implemented with good in-band transmission, improved out-of-band attenuation, and excited notch band within the UWB passband.

5.2 AN INITIAL MMR-BASED UWB BANDPASS FILTER

By integrating the three-mode MMR with two-pole parallel-coupled lines as investigated in Chapter 4, the first three resonant modes of an improved MMR were excited and used to realize five transmission poles with high-return loss in the whole passband. Following the works in References 3 and 4, the MMR with $\theta_2 = \theta_1$ is properly modified in configuration so as to reallocate its first three resonant modes close to the lower end, center, and upper end of the specified UWB passband with reference to the mode graph plotted in Figure 5.1. By largely increasing the coupling degree of the two parallel-coupled lines, a good UWB passband with five transmission poles is realized. In this case, the first- and third-order resonant frequencies basically determine the lower and upper cutoff frequencies of a wide passband, as discussed in Section 4.4.2.

The first MMR-based UWB bandpass filter was reported in Reference 2 on the Rogers RT/Duriod 6010, with the dielectric substrate of $\varepsilon_r = 10.8$ and thickness $h = 1.27$ mm. Figure 5.2 shows its basic schematic layout. At center frequency of the UWB passband, that is, 6.85 GHz, this MMR consists of a SIR with $R_Z > 1$ shown in Figure 4.17b with one half-wavelength ($\lambda_g/2$) low-impedance line section at the center and two identical $\lambda_g/4$ high-impedance line sections at two sides. As shown in Figure 5.2, the strip widths of two identical high-impedance lines at two sides of this MMR are selected as 0.10 mm,

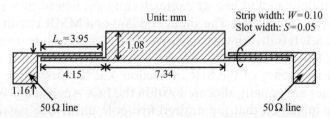

Figure 5.2. Schematic of a MMR-based microstrip-line UWB bandpass filter.

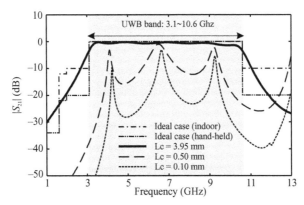

Figure 5.3. Insertion loss of the microstrip-line UWB bandpass filter with varied parallel-coupled line lengths (L_C).

while their lengths are reasonably stretched to about $\lambda_g/4$ at 6.85 GHz. The central low-impedance line portion of this MMR is 1.08 mm in width and approximately $\lambda_g/2$ or 7.34 mm in length. In order to have the freedom to enhance the coupling degree as inquired later on, the strip and slot widths in the coupled lines are chosen as 0.10 and 0.05 mm, respectively. In the full-wave simulation, a strip conductor thickness of 17 μm is used. In order to quantitatively investigate the multiple-mode resonance behaviors of this MMR, the coupled-line length (L_c) is first selected much shorter than $\lambda_g/4$ with respect to 6.85 GHz.

Figure 5.3 plots the simulated S_{21}-magitude graphs in a wide frequency range (1.0 to 13.0 GHz) under three different coupling lengths, that is, L_c = 0.1, 0.5 and 3.95 mm. It needs to be pointed out that the two dash-dot lines in Figure 2 indicate the desired insertion loss configurations of the ideal UWB filters for the indoor and handheld UWB systems [5], respectively. Looking at the two weak coupling cases with L_c = 0.1 and 0.5 mm, the three resonant frequencies with the S_{21}-magnitude peaks are observed to occur at around 4.23, 6.66, and 9.26 GHz, respectively. As L_c increases from 0.1 to 0.5 mm, the S_{21}-magnitude curve slightly rises. As L_c largely increases to 3.95 mm, which approximately equals to $\lambda_g/4$ at 6.85 GHz, the whole S_{21}-magnitude realizes an almost flat frequency response near the 0-dB horizontal line over the desired UWB band. Moreover, the overall length of this UWB filter is approximately equal to one full wavelength at 6.85 GHz, which is much smaller than those reported in References 6–11 using other technologies.

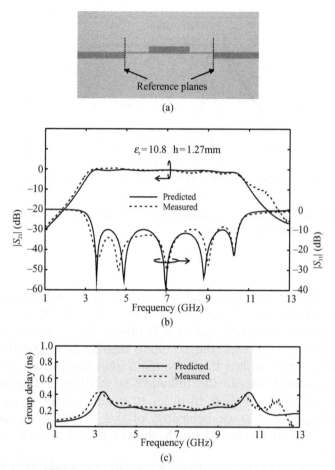

Figure 5.4. Predicted and measured results of the designed microstrip-line UWB band-pass filter. (a) Photograph. (b) S-magnitudes.. (c) Group delay.

Figure 5.4a is the photograph of the fabricated UWB bandpass filter. The length between the two reference planes is about 16 mm. The simulated and measured frequency responses are plotted in Figure 5.4b,c for quantitative comparison. In the wide frequency range of 1.0–13.0 GHz, the measured results are in good agreement with those from simulation, achieving $FBW \approx 113\%$. At 6.85 GHz, the measured insertion loss is found as 0.55 dB, which is much better than 6.7 dB in Reference 7. Over the UWB passband, the return loss in the simulation and measurement are both higher than 10 dB with the appearance of five transmission poles. It can be understood from Figures 4.17 and 4.22 that three poles are brought out by the first three resonant modes of the MMR and two poles are contributed by the two quarter-wavelength parallel-coupled microstrip lines. Within the passband, the simulated and measured group delay are both less than 0.43 ns, with a maximum variation of 0.23 ns, thus implying a good linearity of this designed UWB bandpass filter.

In general, this MMR-based UWB bandpass filter can be realized by utilizing other resonant modes, such as the first two, three, or four resonant modes of a constituted MMR. By varying the length of the center low-impedance line section or increasing the number of nonuniform sections in the MMR, the UWB bandpass filters with increased in-band transmission poles were presented and implemented in References 12–14. In Chapter 6, a direct synthesis approach will be presented for alternative analysis and design of these filters in an efficient manner.

5.3   UWB BANDPASS FILTERS WITH VARIED GEOMETRIES

5.3.1   Aperture-Backed PCML

As discussed in Section 4.2.2, an aperture-back parallel-coupled microstrip line is formed by the partial removal of the ground plane underneath the coupled strips, and it is used herein as a coupling-enhanced parallel-coupled line in the design of an alternative MMR-based UWB filter with relaxed fabrication tolerance. Following the discussion in the previous section, an aperture-backed parallel-coupled line is designed with its maximum coupling strength near the UWB center frequency, that is, 6.85 GHz. By driving a modified three-mode MMR with these parallel-coupled lines at its two sides, a compact UWB bandpass filter with five transmission poles was developed in Reference 15. The physical configuration of this filter is shown in Figure 5.5.

Compared with the parallel-coupled line sections in Figure 5.2, the strip and slot widths of the coupled-line section are increased from 0.1 and 0.05 mm to 0.2 and 0.1 mm with the help of a backside aperture

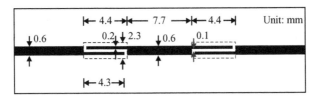

Figure 5.5. Schematic of the designed UWB bandpass filter on an aperture-backed microstrip line.

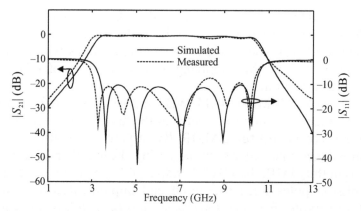

Figure 5.6. Simulated and measured results of the UWB bandpass filter on aperture-backed microstrip line.

underneath the coupled strip conductors. Thus, this approach can tremendously relax the requirement in fabrication tolerance and also reduce the conductor loss as the strip conductor is widened. Figure 5.6 shows the simulated and measured S-parameter magnitudes of the designed UWB bandpass filter. The two sets of results agree reasonably well with each other except that the lower cut-off frequency in the measurement is slightly shifted down to 2.9 GHz. Due to the very small interaction between the fields in the aperture and strip conductor, we find that the radiation-related loss is very small in the concerned UWB. Similar to the filter in Figure 5.2, the overall length of this aperture-compensated UWB filter is only about one guided-wavelength at 6.85 GHz, which is much smaller than those reported in References 6–11.

5.3.2 Broadside Microstrip/CPW

It is well known that the surface-to-surface or broadside coupled structures can also be utilized to enhance the desired coupling strength by employing both the top and bottom surfaces of a substrate. Therefore, a UWB bandpass filter is constructed using a hybrid microstrip/CPW structure with the back-to-back transition configuration, and its physical schematic is shown in Figure 5.7 [16]. Here, an MMR-based CPW resonator is formed on the ground plane with its first three resonant modes, which are allocated in the desired UWB passband. In this aspect, the broadside-coupled microstrip-to-CPW transition or coupling struc-

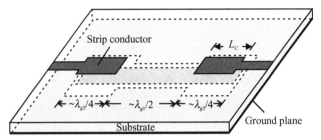

Figure 5.7. Three-dimensional view of proposed UWB bandpass filter on hybrid microstrip/CPW structure.

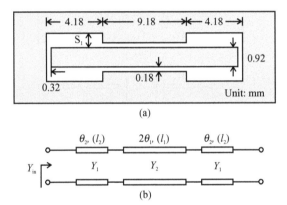

Figure 5.8. Proposed MMR on coplanar waveguide. (a) Layout. (b) Equivalent transmission line network.

ture with enhanced coupling strength is used to replace the parallel-coupled line sections in the initial UWB filter in Figure 5.2.

As shown in Figure 5.8a, the proposed open-ended MMR on CPW is composed of one central CPW with narrow slot width or low impedance and two identical CPWs with wide slot width or high impedance at two sides under the fixed strip width. Figure 5.8b depicts its equivalent transmission line topology with three cascaded sections. To determine the frequencies of the three resonant modes, the input admittance (Y_{in}) at one of the open ends, looking into the MMR, must be zero, $Y_{in} = 0$. Thus, all those resonant frequencies can be solved from a transcendental Equation 4.26.

Table 5.1 tabulates the first three resonant frequencies (f_1, f_2, f_3) versus slot width (S_1) under the fixed slot width of 0.18 mm in the middle. As shown in Figure 5.8a, the three sections in this MMR are selected in such a way that the middle section has about one-half

TABLE 5.1   First Three Resonant Frequencies (f_1, f_2, f_3) versus Slot Width (S_1) for the MMR in Figure 5.8

S_1 (mm)	0.18	0.58	0.98	1.10	1.38	1.78
f_1 (GHz)	3.67	4.01	4.12	4.13	4.15	4.16
f_2 (GHz)	7.29	7.10	6.84	6.76	6.60	6.34
f_3 (GHz)	10.91	10.12	9.55	9.40	9.13	8.74

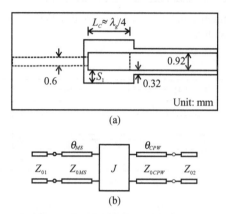

(a)

(b)

Figure 5.9. Broadside microstrip/CPW coupling structure. (a) Layout. (b) Equivalent J-inverter network.

guided wavelength or $L_1 \approx \lambda_{g1}/2 = 9.18$ mm, and the two side sections have about one-quarter guided wavelength or $L_2 \approx \lambda_{g2}/4 = 4.18$ mm at 6.85 GHz, respectively. As S_1 is widened, the first resonant frequency (f_1) quickly rises up at the beginning and then becomes saturated around 4.12–4.16 GHz. On the other hand, the second and third ones (f_2 and f_3) seem to quasi-linearly fall down with S_1.

To achieve a UWB passband covering 3.1 to 10.6 GHz, the first three frequencies are equally spaced in the UWB band with the locations above 3.1 GHz, near 6.85 GHz, and below 10.6 GHz. According to this criteria, the three resonant frequencies of 4.12, 6.84, and 9.55 GHz under $S_1 = 0.98$ mm can be recognized as the best in all the cases listed in Table 5.1.

Figure 5.9a shows a hybrid microstrip/CPW surface-to-surface coupling structure that was initially studied in Reference 17 to make up a broadband microstrip-to-CPW transition with the use of its frequency-dispersive and enhanced coupling extent. In this structure, the upper microstrip conductor is vertically coupled with the central strip conductor of the lower CPW on ground plane via electromagnetic coupling. Its coupling behavior can be characterized in terms of an equivalent

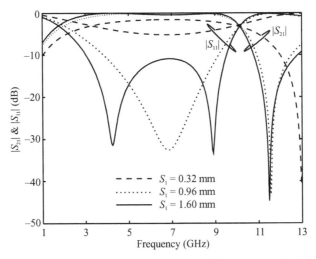

Figure 5.10. Simulated S-parameters of the coupling structure in Figure 5.9.

unified J-inverter network as illustrated in Figure 5.9b. The J-inverter admittance in fact represents the coupling extent and its maximum peak is properly allocated near 6.85 GHz by selecting the coupling length close to $L_C \approx \lambda_g/4$ as shown in Figure 5.9a. Figure 5.10 depicts the simulated frequency-dependent parameters of this coupling structure under different slot widths. Similar to the case of parallel-coupled line in Figure 4.22, the two transmission poles can also be excited in the two sides of 6.85 GHz as S_1 is widened.

As a result, a five-pole UWB bandpass filter is constructed based on the above resonator and coupling structures. The proposed coupled-line structure can address two problematic issues existing in the initial MMR-based UWB filter in Figure 5.2: insufficiently strong coupling strength between the two side-to-side coupled lines and undesired radiation loss from a wide strip conductor in the center, especially at high frequencies. Figure 5.11a shows the top- and bottom-view photographs of the filter fabricated on RT/Duriod 6010 with $\varepsilon_r = 10.8$ and $h = 0.635$ mm. Similarly, only one wavelength is needed in this five-pole filter, and the measured results are shown in Figure 5.11b. Over a wide frequency range, flat insertion loss and group delay are observed as well.

5.3.3   Inductive Coupling Element on CPW

A parallel-coupled CPW with short-circuited ends is used here as an inductive coupling structure that can avoid high radiation loss in the

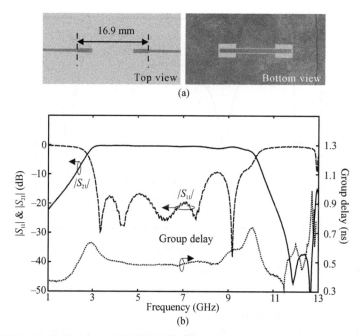

Figure 5.11. A hybrid microstrip/CPW UWB bandpass filter. (a) Photograph. (b) Measurement results.

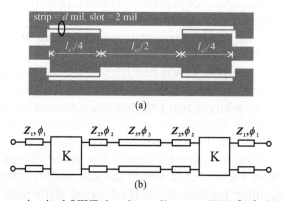

Figure 5.12. Short-circuited UWB bandpass filter on CPW [30]. (a) Geometry. (b) Equivalent circuit network.

open-circuited ends of the conventional capacitively parallel-coupled CPW [18]. Figure 5.12a is the geometry of an alternative UWB bandpass filter based on a short-circuited MMR on CPW and inductively parallel-coupled CPW structure [19]. In this way, a UWB filter can be implemented in a purely CPW topology that is required in direct integration with other uniplanar circuits.

Generally speaking, an asymmetrical parallel-coupled CPW with short-circuited ends can be expressed as a K-inverter network with the impedance (K) and two unequal line lengths (ϕ_1 and ϕ_2) at two sides, as discussed in Reference 20. Therefore, the equivalent network topology of this filter can be described as a cascaded network, as shown in Figure 5.12b. Here, three cascaded sections, composed of the three phases, ϕ_2, ϕ_3, and ϕ_2, located between the two K-inverters, make up the resultant short-circuited MMR.

As compared with the microstrip line, the CPW has an advantageous feature, that is, the strip conductor and ground plane can be easily linked together to make up a short-circuited end. This work presents a short-circuited MMR on CPW whose geometrical sketch is illustrated in Figure 5.13a. As discussed in Section 4.4.2, this short-circuited MMR is composed of three distinctive CPW sections with one high-impedance section in the middle and two identical low-impedance sections at the two sides. Figure 5.13b depicts its equivalent transmission line network, in which the two CPW step discontinuities are negligibly small and they are ignored so as to simplify our following analysis.

To manifest its MMR characteristics for the UWB filter design, the longitudinal resonant condition for all the resonant modes needs to be established. For this purpose, the input impedance at the left short end, looking into the right end, is indicated in its equivalent transmission line network, as shown in Figure 5.13b, such that

$$Z_{\mathrm{in}} = jZ_2 \frac{2(R_Y - \tan\theta_1 \tan\theta_2)(R_Y \tan\theta_1 + \tan\theta_2)}{K(1 - \tan^2\theta_1)(1 - \tan^2\theta_2) - 2(1 + K^2)\tan\theta_1 \tan\theta_2}. \quad (5.1)$$

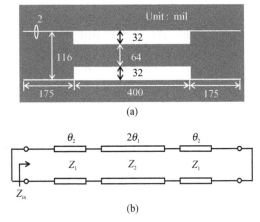

(a)

(b)

Figure 5.13. (a) Geometry and (b) equivalent circuit network of the short-ended CPW MMR.

where $R_Y = Z_1/Z_2$ is the impedance ratio of the middle and end CPW sections in this MMR.

Under the resonant condition of $Z_{in} = 0$, a set of algebraic equations can be formed to determine all the resonant frequencies:

$$R_Y - \tan\theta_1 \tan\theta_2 = 0 \quad \text{(at odd-mode frequencies)} \quad (5.2a)$$

$$R_Y \tan\theta_1 + \tan\theta_2 = 0 \quad \text{(at even-mode frequencies).} \quad (5.2b)$$

As we expected, Equation (5.2) is exactly the same as Equation (4.30). As the lengths of these three sections are readily selected as $\theta_2 = \theta_1 = \theta$, f_1, f_2, and f_3 can be derived from the three closed-form formulas:

$$\theta(f_1) = \tan^{-1}\sqrt{R_Y} \quad (5.3a)$$

$$\theta(f_2) = \frac{\pi}{2} \quad (5.3b)$$

$$\theta(f_3) = \pi - \tan^{-1}\sqrt{R_Y}. \quad (5.3c)$$

In this way, we find that the lower and higher frequencies, f_1 and f_3, are mainly determined by the impedance ratio R_Y, while the central f_2 is somehow affected by the actual lengths of three sections. To realize a UWB passband later on, the traverse slot/strip widths of the middle and end CPW sections should be properly chosen to allocate these three frequencies, f_1, f_2, and f_3 toward the locations of the lower-end, center, and higher-end frequencies of such a UWB passband.

After all the dimensions of this MMR are determined, the coupled spacing (d) is properly adjusted with the target of minimizing the insertion loss within the UWB passband. Figure 5.14 depicts the four separate curves of frequency response of S_{21}-magnitude under $d = 15, 10, 5,$ and 2 mil (the best case). At $d = 15$ mil, the three maximum S_{21}-magnitude peaks are located at 4.2, 7.0, and 9.8 GHz, respectively, while the poor transmission with high insertion loss is observed in the frequency range away from these resonant frequencies. As d is reduced, the coupling degree of the parallel-coupled CPW gains a gradual increment in extent so as to raise the S_{21}-magnitude and reach its maximum 0-dB line at $d = 2$ mil. Moreover, as the coupling is enhanced, the lower and higher cut-off frequencies are found to shift down and up simultaneously to 3.1 GHz and 10.6 GHz, thus achieving a good UWB passband at $d = 2$ mil.

To verify the predicted electrical performance based on the simple network in Figure 5.12b, the optimized UWB filter ($d = 2$ mil) is simulated again over its entire layout using the Agilent Momentum soft-

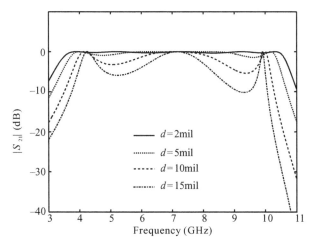

Figure 5.14. Simulated S-magnitudes of the UWB filter in Figure 5.12a under varied d in the coupled CPW.

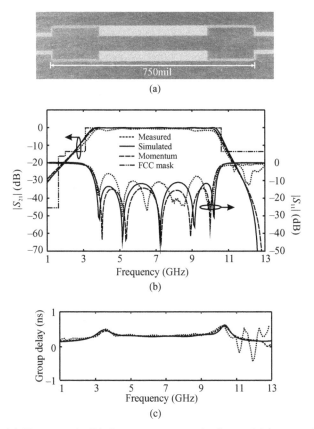

Figure 5.15. (a) Photograph, (b) S-parameter magnitudes, and (c) group delay of predicted, simulated, and measured results of the optimized UWB filter.

ware, and its circuit sample is then fabricated on the Duroid 6010 substrate with the thickness of $h = 25$ mil and permittivity of $\varepsilon_r = 10.8$. The designed UWB filter has the overall length of 750 mil, which is about one guided wavelength at 6.85 GHz. Figure 5.15a shows the photograph of the fabricated circuit. Figure 5.15b,c are the results from the calculation of the cascaded network in Figure 5.12b, direct electromagnetic simulation, and microwave measurement. In the realized UWB passband of 3.3–10.4 GHz, the measured maximum $|S_{21}|$ achieves 1.5 dB as compared with 0.2 dB in the simulation. It is mainly due to unexpected radiation loss in the CPW structure and slightly poor impedance matching.

5.4   UWB FILTERS WITH IMPROVED OUT-OF-BAND PERFORMANCE

5.4.1   Capacitive-Ended Interdigital Coupled Lines

In the design of a UWB bandpass filter, it is a general approach that a MMR resonator is constructed by equally allocating the first three resonant frequencies in the core UWB passband. Meanwhile, the fourth or other higher-order resonant frequencies may contribute spurious passbands, thus degrading the upper stopband performance of a UWB filter shown in Figure 5.16.

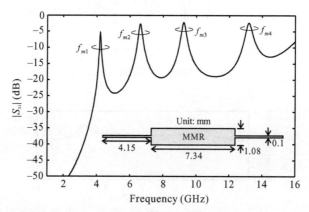

Figure 5.16. Frequency-dependent transmission responses of the constituted initial MMR (as used in Figure 5.2) on the RT/Duroid 6010 with $\varepsilon_r = 10.8$ and $h = 1.27$ mm under a weak coupled excitation.

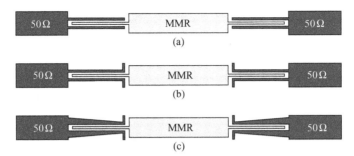

Figure 5.17. Schematics of the three MMR-based UWB bandpass filter. (a) Filter A. (b) Filter B. (c) Filter C.

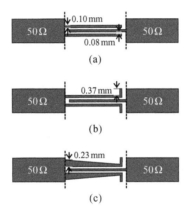

Figure 5.18. Layouts of the three interdigital microstrip coupled lines. (a) Type A. (b) Type B. (c) Type C.

To circumvent this problem, an interdigital coupled line with capacitive-ended loading and/or tapered strip shape is constructed, and its first transmission zero is reallocated toward full suppression of this fourth resonant frequency in the MMR [21]. Figure 5.17a–c depict the schematics of the three UWB filters to be considered, namely, filter A, filter B, and filter C. The interdigital coupled lines with enhanced degree of coupling are used to make up the filter A with increased return loss in the UWB passband, and their modified counterparts with capacitive-ended loading and/or tapered strip shape are then formed to reallocate their first transmission zero to f_{m4}, as discussed in Reference 22, so as to make up the filter B and filter C, respectively.

Figure 5.18a–c further highlight the three types of microstrip inter-digital or double-parallel coupled lines, namely type A, type B, and type C. The type A structure is expected to achieve much tighter coupling

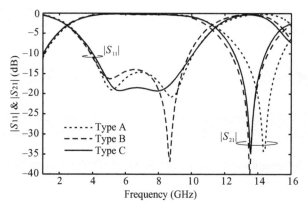

Figure 5.19. Predicted frequency responses of the three interdigital coupled-lines in Figure 5.18.

degree than the conventional parallel coupled line used in the initial UWB filter. The latter two structures with capacitive-ended loading are formed to allocate their coupling zero to the fourth-order resonant frequency, f_{m4}. Figure 5.19 plots the frequency-dependent scattering parameters of these coupling structures via Agilent ADS software. As compared with the coupled line of type A, the coupling zero in the type B moves down from 14.4 to 13.5 GHz as the bottom and top strip arms are capacitively terminated by open-ended stubs. To achieve a balanced UWB passband behavior, the two strip arms with tapered configuration are properly formed in the type C structure. In this aspect, the two transmission poles within the passband can be compensated in a balanced way while the coupling zero keeps at 13.5 GHz, as illustrated in Figure 5.19.

Figure 5.20 shows the schematic and frequency responses of this improved UWB filter. This proposed filter is capable to reduce the insertion loss within the UWB passband and widen the upper-stopband bandwidth as confirmed in experiment. As shown in Figure 5.20, the measured return and insertion losses are higher than 14 dB and lower than 1.3 dB, respectively, over the UWB passband. The measured group delay varies between 0.20 and 0.30 ns in the simulation.

To further sharpen the roll-off skirts near the lower and upper cutoff frequencies, a two-stage UWB bandpass filter is designed with two cascaded MMRs. Figure 5.21a is the photograph of a fabricated UWB bandpass filter with two MMR resonators, which are coupled together via a parallel-coupled line. Figure 5.21b is the predicted and measured results. In the UWB passband, the two return losses keep higher than

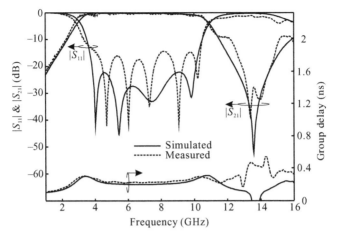

Figure 5.20. Predicted and measured results of the single-stage UWB bandpass filter.

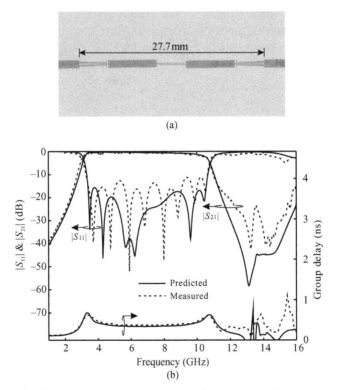

Figure 5.21. Predicted and measured results of the two-stage UWB bandpass filter.

14 dB, and the group delay varies between 0.35 and 0.65 ns with a maximum variation of 0.30 ns. Outside the UWB passband, the lower- and upper-band skirts get sharpened to a great extent while the upper-stopband with the insertion loss above 20 dB occupies an enlarged range of 11.8 to 15.9 GHz.

5.4.2 Coupling between Two Feed Lines

Another approach to enhance the filter selectivity is to introduce extra coupling between the two feed lines [23]. In Reference 24, a UWB bandpass filter is presented using an MMR-based slotline resonator and back-to-back microstrip-slotline transition. This slotline resonator is further twisted into a W-shape, as shown in Figure 5.22 [24], in order to make the overall size smaller and create an additional coupling path. Without increasing any size, two lowpass filters are installed in the sections of two feed lines so as to improve the out-of-band performance.

Figure 5.23a shows the initial geometrical layout of the slotline MMR. It consists of a half wavelength high-impedance (Z_2) section in the middle and the two quarter wavelength low-impedance (Z_1) sections in the two sides. The resonant frequencies of this short-circuited MMR are determined under transmission-line resonances, and they are expressed as the solutions in Equation (5.2). To simplify this study, we

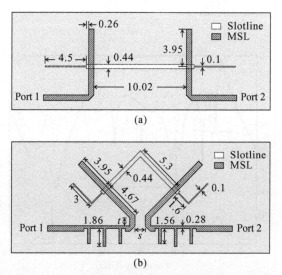

(a)

(b)

Figure 5.22. Schematic layouts of the proposed UWB bandpass filters. (a) Basic structure. (b) W-shaped structure. All units are in mm. Substrate: $\varepsilon_r = 10.8$ and $h = 0.635$ mm.

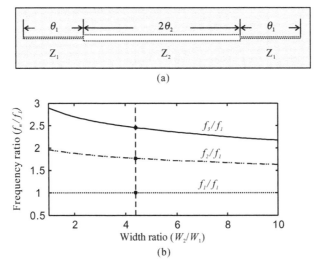

Figure 5.23. (a) Schematic layout. (b) Resonant frequency ratios versus the width ratio of the slotline MMR.

assume $\theta_1 \approx \theta_2$ so that the third resonant frequency over its first counterpart can be derived as Equation (5.3), and it can be expressed as a closed-form formula:

$$\frac{f_3}{f_1} = \frac{\theta(f_3)}{\theta(f_1)} = \frac{\pi}{\tan^{-1}\sqrt{R_Y}} - 1. \tag{5.3}$$

In order to equally allocate the first three resonances close to the lower-end, center, and higher-end frequencies of the targeted UWB passband, the frequency ratio f_3/f_1 should be selected according to the position of pole frequency in Equation (4.33). It can be realized by setting the impedance ratio R_Y slightly larger than 1, as discussed in Section 4.4.2. Figure 4.5b plots the two frequency ratios, that is, f_2/f_1 and f_3/f_1, versus the width ratio (W_2/W_1). In the design, the width (W_1) is selected as its minimum tolerable value of 0.1 mm in existing fabrication procedure while minimizing unexpected conductor loss. As the width (W_2) is gradually raised from 0.1 to 1.0 mm, the first three resonances can be observed to move closer to each other due to enlarged W_2/W_1. When W_2/W_1 is chosen as 4.4, the first three resonant frequencies appear at 4.01, 7.10, and 9.84 GHz, respectively. If two microstrip-to-slotline transitions are properly introduced at the input and output ports, a dominant passband covering the FCC-defined UWB can be eventually constructed, as will be demonstrated later on.

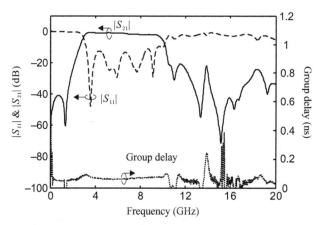

Figure 5.24. Measured S-parameters and group delay of the UWB bandpass filter on microstrip/slotline transition.

The UWB bandpass filter shown in Figure 5.22b is implemented on a substrate with dielectric constant of $\varepsilon_r = 10.8$ and $h = 0.635$ mm. The measured frequency response and group delay are plotted in Figure 5.24, which exhibits an ultra-wide passband from 2.95 to 10.16 GHz or 110% fractional bandwidth at 6.56 GHz. The mid-band insertion loss is 1.27 dB, and the return loss is higher than 9.47 dB within the whole passband. The in-band group delay is less than 0.98 ns with a maximum variation of 0.41 ns. The upper stopband is stretched up to 20 GHz as expected. On the other hand, an extra transmission zero is excited around 1.5 GHz below the UWB passband due to the induced coupling between the two feed lines. This design allows us to implement such a UWB filter on a very thin substrate with relaxed fabrication tolerance.

5.4.3 Stub-Loaded MMR

In order to achieve a wider upper stopband, spurious passband occurring at the fourth resonant frequency becomes our primary concern. Although it can be eliminated by some means as discussed in the above sections, the fourth resonant mode can be utilized together with the first three resonant modes to make up a four-pole passband. As we know, the even-order resonant frequencies can be changed by loading an open-ended stub in shunt at the central plane of the line resonator while the odd-order modes do not receive any influence. At an approximately one-third wavelength at central frequency from both ends, an equivalent perfect magnetic wall or open circuit is valid for the third

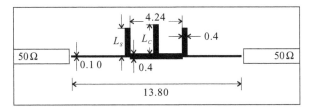

Figure 5.25. Schematic of the novel stub-loaded MMR.

resonant mode. So the third resonant frequency will be significantly affected if an open-ended stub is loaded in shunt at this position. In this aspect, the first four resonant frequencies can be allocated evenly within the UWB passband by two distinct stubs that are placed at two different locations along the resonator. With this idea in mind, the stub-loaded MMR is proposed, and its schematic layout is depicted in Figure 5.25. Generally speaking, as the open-ended stubs are attached in shunt with a uniform transmission-line resonator, effective phase constant or slow-wave factor will be gradually raised in each of transmission-line section, resulting to move forward all the resonant frequencies in a different way.

To provide a quantitative view on the effect of these loaded stubs, Figure 5.26 is prepared to demonstrate the variation of the first four resonant frequencies as a function of central stub length (L_C) and side stub length (L_S). First, in order to study the effect of the central stub on even-order resonant frequencies, L_C is increased gradually from 1.8 to 2.3 and 2.8 mm under the fixed L_S. As can be observed from Figure 5.26a, the second and fourth resonant frequencies fall down with L_C slowly and quickly, while the first and third ones are almost unchanged. On the other hand, by fixing $L_C = 2.29$ mm and changing L_S from 1.7 to 2.2 and 2.7 mm, the third resonant frequency can be gradually reduced from 9.47 to 8.76 and 8.03 GHz as depicted in Figure 5.26b. As a result, the second, third, and fourth resonances can be fully tuned and controlled by the lengths of these stubs placed at two distinct positions along a uniform resonator. Thereafter, by proper choice of these stubs in locations and lengths, the first four resonant modes can be reallocated to make their contribution on the formulation of a UWB passband with six transmission poles.

As discussed earlier, in order to realize such a UWB passband using the stub-loaded resonator in Figure 5.25, two identical frequency-dispersive coupled lines with strong coupling degree need to be simultaneously placed at the two sides of the constituted resonator. Following the coupling structure used in Section 5.4.1, the interdigital coupled

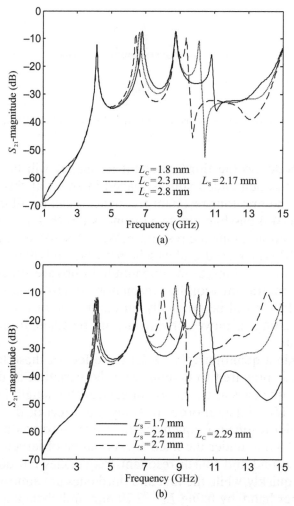

Figure 5.26. Frequency-dependent $|S_{21}|$ of the stub-loaded MMR versus lengths of the loaded stubs. (a) Central-stub length L_C. (b) Side-stub length L_S.

lines are selected, and they are installed in the two sides of the proposed stub-loaded MMR. Since the fourth resonant mode is in addition introduced in the UWB passband, the requirement on high coupling degree of coupled lines is relaxed to a large extent. In this case, the coupling gaps between the parallel lines are all widened to 0.1 mm, which is twice of that in the initial MMR-based bandpass filter in Figure 5.2, thus significantly circumventing the tolerance in fabrication. The configuration of the finally designed UWB bandpass filter with the dimensions labeled is shown in Figure 5.27 [25].

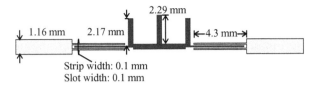

Figure 5.27. Schematic layout of the proposed stub-loaded MMR UWB bandpass filter. Substrate: Roger's RT/Duriod 6010, $\varepsilon_r = 10.8$, and $h = 1.27$ mm.

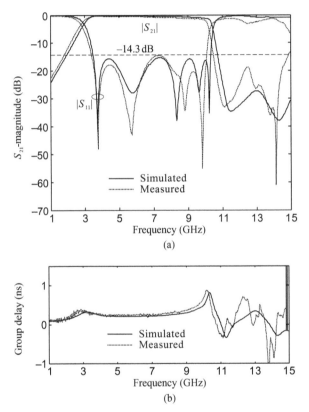

Figure 5.28. Simulated and measured results of stub-loaded MMR-based UWB bandpass filter. (a) S-parameters. (b) Group delays.

Figure 5.28 shows the simulated and measured results of the designed UWB bandpass filter based on the stub-loaded MMR. The lower and higher cut-off frequencies of the constituted UWB passband are equal to 2.80 and 10.27 GHz in experiment compared with their counterparts of 2.98 and 10.49 GHz in simulation. The fabricated filter achieves a return loss higher than 14.3 dB over the realized UWB passband with a fractional bandwidth of 114%. As compared with the initial MMR-based UWB bandpass filter in Figure 5.4, this filter gets the increased roll-off in the upper cutoff frequency due to simultaneous emergence of four resonant modes occurring in the UWB passband. In the meantime, both simulated and measured group delays within the UWB passband are found less than 0.87 and 0.81 ns, respectively, with the maximum variation of 0.64 ns. This raised variation is caused by a trade-off between the rejection skirt in S_{21}-magnitude and linearity in phase as concluded in Reference 26. In final, the overall length of the proposed UWB bandpass filter is 13.80 mm, and it is shorter than 15.64 mm of the initial filter in Figure 5.2. It is primarily benefited by enhanced slow-wave factor of stub-loaded structure involved in the modified MMR.

5.4.4 Lowpass Embedded MMR

To suppress the undesired harmonic passbands beyond the core UWB passband, various low-pass structures with high-frequency rejection can be externally installed with the UWB filter as shown in one filter example in Figure 5.22b. In order to improve the upper-stopband performance without enlargement of the overall size of a filter circuit, it is always preferred to internally embed low-pass structures within the MMR itself. In this aspect, the proposed filter topology should be composed of a stopband-contained resonator and two parallel-coupled lines. Figure 5.29 shows one filter example based on this design methodology. Two capacitive-ended coupled lines with lowpass properties are symmetrically arranged at the two sides in a MMR. In order to get a tight coupling degree while keeping a low conductor loss, the two wide apertures are placed on ground plane underneath two coupled lines with the slot/strip widths of 0.10 mm.

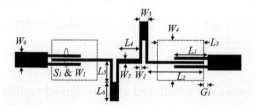

Figure 5.29. Schematic layout of the proposed low-pass embedded MMR UWB bandpass filter. Substrate: Roger's RT/Duriod 6010, $\varepsilon_r = 10.8$ and $h = 0.635$ mm. $L_1 = 4.0$, $L_2 = 4.85$, $L_3 = 3.90$, $L_4 = 1.02$, $L_5 = 1.05$, $L_6 = 1.0$, $W_1 = 0.10$, $W_2 = 0.10$, $W_3 = 0.20$, $W_4 = 1.70$, $W_5 = 0.40$, $W_6 = 0.6$, and $G_1 = 0.10$. All dimensions are given in millimeter (mm).

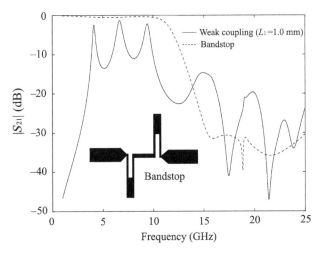

Figure 5.30. Simulated S_{21}-magnitude of a stopband-embedded resonator under the weak coupling (solid line) and the stopband section with the 50-Ω feed lines (dotted line).

Figure 5.30 shows the simulated S-parameter magnitudes (solid line) under weak coupling with length $L_1 = 1.0$ mm. Under the weak coupling, the first three resonant frequencies are visually located around 4.2, 6.7, and 9.4 GHz within passband, while the other undesired resonant frequencies appear above passband and become harmonics of UWB bandpass filter. On the other hand, the simulated results of the cascade stopband or lowpass sections with the 50-Ω feed lines are shown in Figure 5.30 (dash line). From the results, we can observe that the attenuation of stopband achieves 30 dB up to 25 GHz, and can be further utilized to suppress all undesired harmonics of the constituted UWB bandpass filter.

Figure 5.31a is a photograph of the fabricated UWB filter [27]. As a benefit of three controllable transmission zeros excited in the capacitive-ended coupled lines, an improved UWB filtering performance with a stretched upper-stopband up to 25 GHz is attained, as shown in Figure 5.31b. This filter is also implemented on RT/Duroid 6010 with a substrate thickness of 0.635 mm and a dielectric constant of 10.8. As shown in Figure 5.31b, the measured insertion loss is 1.6 dB at the center of the UWB passband and remains below 2.0 dB from 3.1 to 9.0 GHz. The measured return loss exceeds 13.5 dB over the passband. Experimentally, the 3-dB passband ranges from 3.0 to 10.65 GHz, covering the FCC-specified UWB passband. In addition, the overall length of this filter is reduced to only about 11.52 mm.

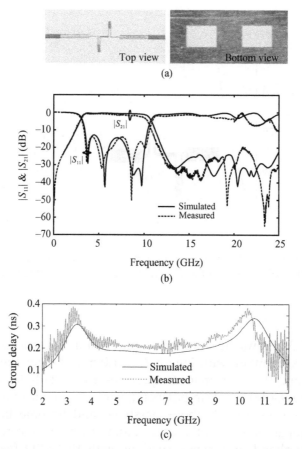

Figure 5.31. Photograph, simulated/measured S-magnitudes, and group delay of the designed UWB filter. Photograph of the fabricated UWB filter (Top and bottom view). (b) S-magnitudes. (c) Group delay.

5.5 UWB BANDPASS FILTER WITH A NOTCH BAND

All of the MMR-based UWB bandpass filters designed above have good in-band performance and are suitable for practical implementation. Nevertheless, the FCC-specified emission mask cannot provide enough protection from harmful interference from generic UWB applications for existing and planned radio systems. Users will have to pay much attention on the exclusive use of the spectrum without causing interference to other existing services, such as wireless local-area network (WLAN) at 5.60 GHz. To circumvent this issue, single [28–34], or multiple [34], narrow-band notched UWB filters were developed. By

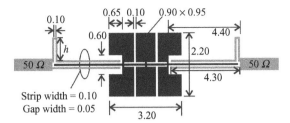

Figure 5.32. Schematic of the proposed band notched UWB bandpass filter (all in mm) [32].

introducing slotline or open-circuited stubs into the filtering topologies, a notch or narrow rejection band can be generated in a certain frequency range in the UWB passband. In References 31 and 32, asymmetric parallel-coupled or interdigital-coupled lines are designed for excitation of a notch band.

In this section, the above-used interdigital coupled lines are rearranged by lengthening and folding one of two coupling arms associated with the feed lines, so as to create a notch band at 5.6 GHz within the UWB passband. As shown in Figure 5.32, two coupling arms now have dissimilar electrical lengths. Thus, the two signal paths along them become out-of-phase and equal magnitude in a certain frequency in which transmission signal is cancelled, thus generating a notch band. This approach is applicable to all UWB filter with interdigital coupling section. To demonstrate this performance, the folded coupling arms and the low-pass embedded resonator are investigated individually, as shown in Figure 5.33. As the length of folded arm varies from 1.60, 1.40 to 1.20 mm, the transmission zero has a gradual increment from 5.25, 5.40 to 5.60 GHz, as shown in Figure 5.33a. With this result, it can be figured out that as the length is set up as $h = 1.60$ mm (solid line), a notch band can be introduced at 5.60 GHz within the UWB passband. In the design of such a notch-band UWB bandpass filter, the spacing and length of an interdigital coupled-line need to be first optimized for minimizing the in-band insertion loss of a UWB bandpass filter. Then, the stretched length ($2h$) of a folded arm is adjusted for allocating a coupling or transmission zero to the notch band at 5.6 GHz.

On the other hand, the stepped impedance stubs are used to be integrated with the MMR and thus largely extend the upper stopband. Figure 5.33b illustrates the circuit layout and simulated S_{21}-magnitude of this MMR circuit under a different coupling length of $d = 0.6$ and 4.3 mm, respectively. As discussed in Section 5.4.3, the first three resonant frequencies, ranged in the 3.1–10.6 GHz band, work together

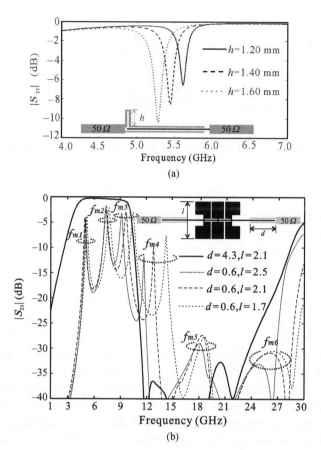

Figure 5.33. S_{21}-magnitudes of extracted interdigital coupling section with varied h values, and of designed MMR with different coupling lengths. (a) Interdigital coupling section. (b) Low-pass embedded MMR.

properly to make up the desired UWB passband. As d increases, the S_{21}-magnitude curve in the UWB band gradually rises up toward the ideal 0-dB line. In the case of $d = 4.3$ mm (solid line), a desired UWB passband is well realized with the bandwidth controlled by the resonant frequencies of the first three resonant modes, that is, f_{m1}, f_{m2}, and f_{m3}. As shown in Figure 5.33b, the third and fourth resonant frequencies, f_{m3} and f_{m4}, gradually increase as the stub length (l) is shortened from 2.5, 2.1 to 1.7 mm. As shown in Figure 5.33b (dotted-line), the fourth-order mode f_{m4} occurs around 12.7 GHz, which is close to the coupling transmission zero of two interdigital coupled lines. This spurious harmonic due to the fourth-order resonant mode is entirely eliminated as d is set as 4.3 mm, as shown in Figure 5.33b (solid line). In addition, the low-

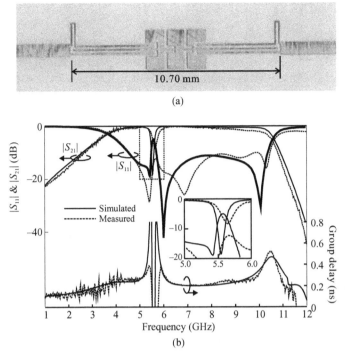

(a)

(b)

Figure 5.34. Photograph, simulated/measured S-magnitudes, and group delay of the designed UWB filter. (a) Photograph of the fabricated UWB filter (top and bottom view). (b) S-magnitudes and group delay.

pass behavior of three paired stubs highly rejects the two spurious passbands centered at around 18.0 and 26.0 GHz, respectively. So, a UWB passband with an extended upper stopband is achieved in the range of 11.0 ~ 26.0 GHz with an insertion loss higher than 20 dB.

By combining the folded coupling arms and the low-pass embedded MMR shown in Figure 5.32, a UWB bandpass filter with a WLAN notch is implemented, and the photograph of its fabricated circuit is shown in Figure 5.34a. With the help of the asymmetrical topology of coupled line arms, a narrow notch band is generated. Simulated and measured frequency responses of this filter are illustrated in Figure 5.34b. Over the plotted frequency range of 1.0–12.0 GHz, the realized UWB passband exhibits the emergence of a notch band at 5.59 GHz with an 18.8-dB rejection and a 3-dB notch bandwidth of 4.6%. In the UWB passband, except for the notch band, the return loss is higher than 11.0 dB, and the group delay is almost flat except for the sharp variation around the edges of such a notch band.

5.6 SUMMARY

In this chapter, various multiple resonant modes (MMRs) with stepped impedance and stub-loaded configurations were extensively investigated and discussed based on transmission line theory. Practically, the MMR-based UWB filters presented exhibit a good UWB passband and compact overall size. To improve the out-of-band performance and avoid interference with other served bands, we presented various UWB bandpass filter designs. All these filters exhibit their unique features, such as an ultrawide passband, a wide upper-stopband, an improved lower-stopband, as well as a WLAN notch band. As highlighted in this chapter, the MMRs have played an indispensable role in the development of such a class of wideband bandpass filters for UWB transmission systems.

REFERENCES

1 L. Zhu, H. Bu, and K. Wu, "Aperture compensation technique for innovative design of ultra-broadband microstrip bandpass filter," *IEEE MTT-S Int. Dig.*, vol. 1 (2000) 315–318.

2 L. Zhu, S. Sun, and W. Menzel, "Ultra-wideband (UWB) bandpass filters using multiple-mode resonator," *IEEE Microw. Wireless Compon. Lett.* 15(11) (2005) 796–798.

3 L. Zhu, W. Menzel, K. Wu, and F. Boegelsack, "Theoretical characterization and experimental verification of a novel compact broadband microstrip bandpass filter," *2001 Asia-Pacific Microwave Conf. Proc.*, December 2001, pp. 625–628, .

4 W. Menzel, L. Zhu, K. Wu, and F. Bogelsack, "On the design of novel compact broad-band planar filters," *IEEE Trans. Microwave Theory & Tech.* 51(2) (2003) 364–370.

5 Federal Communications Commission, "Revision of part 15 of the commission's rules regarding ultra-wideband transmission systems," *Tech. rep., ET-Docket* FCC02-48 (2002) 98–153.

6 H. Ishida and K. Araki, "Design and analysis of UWB bandpass filter with ring filter," *IEEE Topical Conf. on Wireless Comm. Tech.*, 2003, pp. 457–458.

7 A. Saito, H. Harada, and A. Nishikata, "Development of band pass filter for ultra wideband (UWB) communication," *IEEE Conf. Ultra Wideband Systems Tech.* 2003, pp. 76–80.

8 C.-L. Hsu, F.-C. Hsu, and J.-T. Kuo, "Microstrip bandpass filters for ultra-wideband (UWB) wireless communications," *IEEE MTT-S Int. Dig.*, June 2005, pp. 679–682.

9 W. Menzel, M. S. R. Tito, and L. Zhu, "Low-loss ultra-wideband (UWB) filters using suspended stripline," *Proc. 2005 Asia-Pacific Microw. Conf.*, vol. 4, 2005, 2148–2151.

10 R. Gomez-Garcia and J. I. Alonso, "Systematic method for the exact synthesis of ultra-wideband filtering responses using high-pass and low-pass sections," *IEEE Trans. Microw. Theory Tech.* 54(10) (2006) 3751–3764.

11 C.-W. Tang and M.-G. Chen, "A microstrip ultra-wideband bandpass filter with cascaded broadband bandpass and bandstop filters," *IEEE Trans. Microw. Theory Tech.* 55(11) (2007) 2412–2418.

12 Y.-C. Chiou, J.-T. Kuo, and E. Cheng, "Broadband quasi-chebyshev bandpass filters with multimode stepped-impedance resonators (SIRs)," *IEEE Trans. Microw. Theory Tech.* 54(8) (2006) 3352–3358.

13 P. Cai, Z. Ma, X. Guan, Y. Kobayashi, T. Anada, and G. Hagiwara, "Synthesis and realization of novel ultra-wideband bandpass filters using 3/4 wavelength parallel-coupled line resonators," *Proc. Asia–Pacific Microw. Conf.*, December 2006, pp. 159–162.

14 P. Cai, Z. Ma, X. Guan, Y. Kobayashi, T. Anada, and G. Hagiwara, "A novel compact ultra-wideband bandpass filter using a microstrip stepped-impedance four-modes resonator," *IEEE MTT-S Int. Dig.*, June 2007, pp. 751–754.

15 L. Zhu and H. Wang, "Ultra-wideband bandpass filter on aperture-backed microstrip line," *Electron. Lett.* 41(18) (2005) 1015–1016.

16 H. Wang, L. Zhu, and W. Menzel, "Ultra-wideband bandpass filter with hybrid microstrip/CPW structure," *IEEE Microw. Wireless Compon. Lett.* 15(12) (2005) 844–846.

17 L. Zhu and W. Menzel, "Broad-band microstrip-to-CPW transition via frequency-dependent electromagnetic coupling," *IEEE Trans. Microw. Theory Tech.* 52(5) (2004) 1517–1522.

18 L. Zhu, "Realistic equivalent circuit model of coplanar waveguide open circuit: Lossy shunt resonator network," *IEEE Microw. Wireless Compon. Lett.* 12(5) (2002) 175–177.

19 J. Gao, L. Zhu, W. Menzel, and F. Bogelsack, "Short-circuited CPW multiple-mode resonator for ultra-wideband (UWB) bandpass filter," *IEEE Microw. Wireless Compon. Lett.* 16(3) (2006) 104–106.

20 J. Gao and L. Zhu, "Asymmetric parallel-coupled CPW stages for harmonic suppressed $\lambda/4$ bandpass filters," *Electron. Lett.* 40(18) (2004) 1122–1123.

21 S. Sun and L. Zhu, "Capacitive-ended interdigital coupled lines for UWB bandpass filters with improved out-of-band performance," *IEEE Microw. Wireless Compon. Lett.* 16(8) (2006) 440–442.

22 S. Sun and L. Zhu, "Periodically nonuniform coupled microstrip line filters with harmonic suppression using transmission zero reallocation," *IEEE Trans Microwave Theory Tech.* 53(5) (2005) 1817–1822.

23   H. Shaman and J.-S. Hong, "A novel ultra-wideband (UWB) bandpass filter (BPF) with pairs of transmission zeros," *IEEE Microw. Wireless Compon. Lett.* 17(2) (2007) 121–123.

24   R. Li and L. Zhu, "Ultra-wideband microstrip-slotline bandpass filter with enhanced rejection skirts and widened upper stopband," *Electron. Lett.* 43(24) (2007) 1368–1369.

25   R. Li and L. Zhu, "Compact UWB bandpass filter using stub-loaded multiple-mode resonator," *IEEE Microw. Wireless Compon. Lett.* 16(8) (2006) 440–442.

26   J.-S. Hong and M. J. Lancaster, *Microwave Filters for RF/Microwave Applications*, John Wiley & Sons, Inc, New York, 2001.

27   T. B. Lim, S. Sun, and L. Zhu, "Compact ultra-wideband bandpass filter using harmonic-suppressed multiple-mode resonator," *Electron. Lett.* 43(22) (2007) 1205–1206.

28   K. Li, D. Kurita, and T. Matsui, "Dual-band ultra-wideband bandpass filter," *IEEE MTT-S Int. Dig.*, June 2007, pp. 1193–1196.

29   W. Menzel and P. Feil, "Ultra-wideband (UWB) filter with WLAN notch," *36th European Microw. Conf.*, September 2006, 595–598.

30   H. Shaman and J.-S. Hong, "Ultra-Wideband (UWB) bandpass filter with embedded band notch structures," *IEEE Microw. Wireless Compon. Lett.* 17(3) (2007) 193–195.

31   H. Shaman and J.-S. Hong, "Asymmetric parallel-coupled lines for notch implementation in UWB filters," *IEEE Microw. Wireless Compon. Lett.* 17(7) (2007) 516–518.

32   S. W. Wong and L. Zhu, "Implementation of compact UWB bandpass filter with a notch-band," *IEEE Microw. Wireless Compon. Lett.* 18(1) (2008) 10–12.

33   G.-M. Yang, R. Jin, C. Vittoria, V. G. Harris, and N. X. Sun, "Small ultra-wideband (UWB) bandpass filter with notched band," *IEEE Microw. Wireless Compon. Lett.* 18(3)(2008) 176–178.

34   K. Li, D. Kurita, and T. Matsui, "UWB bandpass filters with multi notched bands," *36th European Microw. Conf.*, September 2006, 591–594.

CHAPTER 6

SYNTHESIS APPROACH FOR UWB FILTERS

6.1 INTRODUCTION

The synthesis approach for microwave filters is always highly demanded [1–7] due to its unparalleled high efficiency in systematically determining the overall filter dimensions once the design specifications are given. As discussed in Chapter 3, all the element values for a conventional parallel-coupled line bandpass filter with different orders and ripple levels can be immediately found out from the existing design tables and insertion loss graphs, thus bringing out much convenience for all the designers. Without tedious optimization procedure based on time-consuming full-wave simulation, synthesis approach directly establishes all the element values or parameters in a filter network as a closed-form function of required specifications of a filter. Even though synthesis approach cannot account for frequency dispersion and discontinuities involved in a microwave filter on transmission line topology, it can at least directly determine the initial dimensions of the filter so as not only to significantly reduce the design circles but also to give a clear insight into the working principle of a filter. In this chapter, a direct synthesis approach will be introduced and explained to design a

Microwave Bandpass Filters for Wideband Communications, First Edition. Lei Zhu, Sheng Sun, Rui Li.
© 2012 John Wiley & Sons, Inc. Published 2012 by John Wiley & Sons, Inc.

class of UWB bandpass filters with Chebyshev equal-ripple responses in the concerned UWB passband. The MMR-based UWB filters in Chapter 5 are reinvestigated and redesigned using the synthesis method. The objective of this approach is to provide an alternative point of view and a high-efficient design methodology on these UWB bandpass filters that were initially proposed under the concept of a multiple-mode resonator (MMR).

6.2 TRANSFER FUNCTION

For any synthesis approach, the most critical part is to define and/or find out a transfer function which directly links the transmission line network with the targeted response. In Chapter 3, the maximally flat and Chebyshev equal-ripple functions were presented for synthesis design of a variety of narrow-band bandpass filters. In this chapter, a modified Chebyshev function will be introduced based on an early work in Reference 7 to regulate an equal-ripple response in a wide passband with multiorder transmission zeros at specified locations. First, let us define two variables: $x = \cos\phi = \alpha\cos\theta$ and $y = \cos\xi$. These variables are related to each other as follows:

$$y = x\sqrt{\frac{\alpha^2 - 1}{\alpha^2 - x^2}}, \tag{6.1}$$

and α is a quantity used to determine the bandwidth of a bandpass filter. It is defined as

$$\alpha = \frac{1}{\cos\theta_c}, \tag{6.2}$$

where θ_c is the electrical length at the lower cutoff frequency f_c of the first passband shown in Figure 6.1. As a result, the fractional bandwidth (*FBW*) of the filter can then be defined as

$$FBW = \frac{(\pi - \theta_c) - \theta_c}{\pi/2}. \tag{6.3}$$

In this way, a modified Chebyshev transfer function can be formed as

$$|S_{21}|^2 = \frac{1}{1 + \varepsilon^2 \cos^2(n\phi + q\xi)}, \tag{6.4}$$

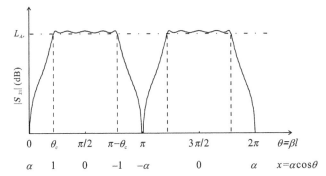

Figure 6.1. Chebyshev bandpass response with high order transmission zeros at $\theta = 0$, π, and 2π ($n = 4$, $q = 1$).

where n and q are integer numbers. The cosine term in the denominator can be expanded as

$$\cos(n\phi + q\xi) = T_n(x)T_q(y) - V_n(x)V_q(y), \tag{6.5}$$

where $T_n(x)$ is the Chebyshev polynomial of the first kind. $V_n(x)$ relates $U_n(x)$, which is the Chebyshev polynomial of the second kind by

$$V_n(x) = \sqrt{1 - x^2}\,U_{n-1}(x). \tag{6.6}$$

By using the relationship between the Chebyshev polynomials of the first kind and the second kind,

$$T_{n+1}(x) = xT_n(x) - (1 - x^2)U_{n-1}(x), \tag{6.7}$$

and substituting Equation (6.6) into Equation (6.7), we have

$$T_{n+1}(x) - xT_n(x) = -\sqrt{1 - x^2}\,V_n(x) \tag{6.8}$$

After replacing all the variables with the functions of θ, we can deduce $|S_{21}|$ as

$$|S_{21}|^2 = \frac{1}{1 + \left|\dfrac{P_{n+q}(\cos\theta)}{\sin^q \theta}\right|^2}, \tag{6.9}$$

where P_{n+q} is an even/odd polynomial function with the highest degree of $(n + q)$. This transfer function presents a bandpass filtering behavior with $(n + q)$ transmission poles and qth order transmission zeros at

$x = \pm\alpha$, as shown in Figure 6.1. It is adopted in a transmission line network with multiple stubs, and the values of n and q primarily depend on the type of stubs inclusive of open- and shorted-circuited stubs in shunt or series. Various combinations of these stubs will be presented later on to make up a variety of UWB bandpass filters.

6.3 TRANSMISSION LINE NETWORK WITH PURE SHUNT/SERIES STUBS

The highpass or quasi-bandpass filter consists of N shunt short-circuited stubs cascaded through $(N-1)$ sections of uniform connecting lines is shown in Figure 6.2 [8], which is an alternative basic structure to implement UWB filters. The electrical length of the stubs and connecting lines are set as θ_c and $2\theta_c$ at a specified cutoff frequency f_c. The normalized impedances for the stubs and connecting lines are indicated by y_k and $y_{k-1,k}$ ($k = 1, 2, \ldots, N$). The first step in the synthesis design method is to establish a transfer function for a given network. Intuitively speaking, a shunt short-circuited stub is equivalent to a highpass element that possess transmission zero at multiple integer times of π, including DC. From Table 2.1, the *ABCD* matrix of such a lossless shunt element can be derived as

$$\begin{bmatrix} 1 & 0 \\ \dfrac{y_i}{j\tan\theta} & 1 \end{bmatrix}, \tag{6.10}$$

and the *ABCD* matrix of the connecting line becomes

$$\begin{bmatrix} \cos 2\theta & j\dfrac{1}{y_{i-1,i}}\sin 2\theta \\ jy_{i-1,i}\sin 2\theta & \cos 2\theta \end{bmatrix}. \tag{6.11}$$

By multiplying the *ABCD* matrixes of the stubs and lines in sequence, the overall *ABCD* matrix of the network in Figure 6.2 can be deduced as

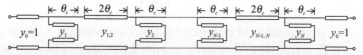

Figure 6.2. Transmission line model of the bandpass filter with shunt stubs.

$$\begin{bmatrix} \bar{A}_{2N-2}(\cos\theta) & j\tan\theta\bar{B}_{2N-2}(\cos\theta) \\ -j\cot\theta\bar{C}_{2N-2}(\cos\theta) & \bar{A}_{2N-2}(\cos\theta) \end{bmatrix}, \tag{6.12}$$

where A_{2N-2}, B_{2N-2}, C_{2N-2}, and D_{2N-2} are the four polynomial functions with degree $(2N-2)$. From Equations (2.32) and (3.9), we know that the magnitude square of S_{21} of a two-port network can be expressed in terms of $ABCD$ of the entire circuit as

$$|S_{21}|^2 = \frac{1}{1+|F|^2}, \tag{6.13}$$

where F is defined as a characteristic function by

$$F = \frac{\Gamma}{T}. \tag{6.14}$$

For a symmetrical network shown in Figure 6.2, F can be simplified as

$$F = \frac{B-C}{2}. \tag{6.15}$$

By substituting Equation (6.12) into Equation (6.15), the $|S_{21}|^2$ of the network in Figure 6.2 is derived as

$$|S_{21}|^2 = \frac{1}{1+\left|j\dfrac{P_{2N-1}(\cos\theta)}{\sin\theta}\right|^2}, \tag{6.16}$$

where P_{2N-1} is a polynomial function with degree $(2N-1)$.

For the next step in this synthesis method, the targeted transfer function Equation (6.9) must be exactly equal to the one in Equation (6.16), which was derived from the network in Figure 6.2, such that

$$\frac{P_{n+q}(\cos\theta)}{\sin^q\theta} = \frac{P_{2N-1}(\cos\theta)}{\sin\theta}, \tag{6.17}$$

where the three integer numbers are related with each other through

$$\begin{aligned} n &= 2N-2 \\ q &= 1. \end{aligned} \tag{6.18}$$

Figure 6.3 shows a dual network of the above-addressed network in Figure 6.2. This dual network is composed of pure series open-circuited stubs with the same electrical length θ_c. Thus, the presented approach

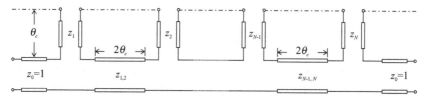

Figure 6.3. Transmission line model of the bandpass filter with series stubs.

can be directly applied to the analysis of this dual network. For the convenience in derivation, normalized impedances are adopted in this network instead of normalized admittances. The *ABCD* matrix of a series open-circuited stub circuit is given as

$$\begin{bmatrix} 1 & \dfrac{z_i}{j\tan\theta} \\ 0 & 1 \end{bmatrix}.$$

(6.19)

By multiplying Equation (6.19) with the *ABCD* matrix of the 2θ transmission line section (Eq. 6.11), the overall *ABCD* matrix of the network in Figure 6.3 is given

$$\begin{bmatrix} \bar{A}_{2N-2}(\cos\theta) & -j\cot\theta\bar{C}_{2N-2}(\cos\theta) \\ j\tan\theta\bar{B}_{2N-2}(\cos\theta) & \bar{A}_{2N-2}(\cos\theta) \end{bmatrix}.$$

(6.20)

By comparing Equation (6.12) with Equation (6.20), it is observed that the two entries A and D are the same, while the other two entries B and C are exchanged with each other. Since the transfer function only deals with the magnitude of the characteristic function F, the network with series stubs shares the same transfer function as the one with shunt stubs. In another way, one can understand that equality of two denominators in Equations (6.9) and (6.16) leads to two individual solutions with the other one besides Equation (6.17) determined by

$$\frac{P_{n+q}(\cos\theta)}{\sin^q\theta} = -\frac{P_{2N-1}(\cos\theta)}{\sin\theta}.$$

(6.21)

Hence, the values of n and q remain the same as those in Equation (6.18).

Furthermore, we notice that the highpass filters shown in Figures 6.2 and 6.3 can be both transformed to lowpass networks by reversing the stub termination type, that is, swapping the short end with the open end and vice versa. So, the insertion loss function of the lowpass filters can

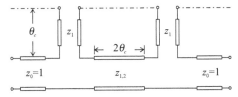

Figure 6.4. Transmission line model of bandpass filter with series stubs and $N = 2$.

be directly derived as what we did previously in the transformation from one to the other highpass structure. In this case, the expressions can be simply obtained by replacing θ in Equation (6.16) with $(\pi/2 - \theta)$, and redefining $x = \alpha \sin\theta$ and $x_c = \sin\theta_c$ in the transfer function (Eq. 6.5). A design example for this type of lowpass filters will be given later on.

A two-stage bandpass filter with series open-ended stubs is shown in Figure 6.4, and it is used as a design example to illustrate how all the element values involved in a filter network are determined using this synthesis method under the predetermined specifications. Due to the topological symmetry of this two-port network, we can have $z_1 = z_2$. The *ABCD* matrix of the entire network can be obtained via multiplication of all the cascaded elements, such that

$$
\begin{bmatrix} A & B \\ C & D \end{bmatrix}
$$
$$
= \begin{bmatrix} 1 & \dfrac{z_1}{j\tan\theta} \\ 0 & 1 \end{bmatrix}
\begin{bmatrix} \cos 2\theta & jz_{1,2}\sin 2\theta \\ j\dfrac{1}{z_{1,2}}\sin 2\theta & \cos 2\theta \end{bmatrix}
\begin{bmatrix} 1 & \dfrac{z_1}{j\tan\theta} \\ 0 & 1 \end{bmatrix}
$$
$$
= \begin{bmatrix} \cos 2\theta + \dfrac{z_1}{z_{1,2}}\cot\theta\sin 2\theta & -j\left(2z_1\cos 2\theta\cot\theta - z_{1,2}\sin 2\theta + \dfrac{z_1^2}{z_{1,2}}\cot^2\theta\sin 2\theta\right) \\ j\dfrac{1}{z_{1,2}}\sin 2\theta & \cos 2\theta + \dfrac{z_1}{z_{1,2}}\cot\theta\sin 2\theta. \end{bmatrix}.
$$
$$
(6.22)
$$

The characteristic function is given by Equation (6.15) as

$$
F = -j\left[\left(2z_1 + z_{1,2} - \frac{1}{z_{1,2}} + \frac{z_1^2}{z_{1,2}}\right)\frac{\cos^3\theta}{\sin\theta} - \left(z_1 + z_{1,2} - \frac{1}{z_{1,2}}\right)\frac{\cos\theta}{\sin\theta}\right]. \quad (6.23)
$$

From Equation (6.18), we can well understand that in order to make these two equations equal to each other, $n = 2$ and $q = 1$ should be adopted. So, Equation (6.9) becomes

$$
\cos(2\phi + \xi) = T_2(x)T_1(y) - V_2(x)V_1(y). \quad (6.24)
$$

The first product term with the Chebyshev polynomial of the first kind are obtained as

$$T_2(x) = 2x^2 - 1 \tag{6.25a}$$

$$T_1(y) = y. \tag{6.25b}$$

From Equation (6.6), $V_2(x)$ and $V_1(y)$ can be derived as

$$V_2(x) = -\frac{1}{\sqrt{1-x^2}}(T_3(x) - xT_2(x)) \tag{6.26a}$$
$$= 2x\sqrt{1-x^2}$$

$$V_1(y) = -\frac{1}{\sqrt{1-y^2}}(T_2(y) - yT_1(y)) \tag{6.26b}$$
$$= \sqrt{1-y^2}.$$

By substituting Equations (6.25) and (6.26) into Equation (6.24), we have

$$\varepsilon\cos(n\phi + q\xi) = 2\varepsilon\left(\frac{1+\sqrt{1-\cos^2\theta_c}}{\cos^3\theta_c}\right)\frac{\cos^3\theta}{\sin\theta} - \varepsilon\left(\frac{2+\sqrt{1-\cos^2\theta_c}}{\cos\theta_c}\right)\frac{\cos\theta}{\sin\theta}. \tag{6.27}$$

Under equalization of the coefficients in Equations (6.23) and (6.27), we deduce

$$\begin{cases} z_1 + z_{1,2} - \dfrac{1}{z_{1,2}} = \varepsilon\left(\dfrac{2+\sqrt{1-\cos^2\theta_c}}{\cos\theta_c}\right) \\[4mm] 2z_1 + z_{1,2} - \dfrac{1}{z_{1,2}} + \dfrac{z_1^2}{z_{1,2}} = 2\varepsilon\left(\dfrac{1+\sqrt{1-\cos^2\theta_c}}{\cos^3\theta_c}\right) \end{cases}. \tag{6.28}$$

Now, we can determine all the normalized characteristic impedances involved in the filter network if the specified ripple level ε and the cutoff-frequency phase θ_c or FBW are known. Tables 6.1–6.4 tabulate the typical element values of the network in Figure 6.3 for $N = 2$ and 3 versus varied θ_c under the specified ripple level of 0.1 and 1.0 dB, respectively. Note that the element values for the filter with shunt short-circuited stubs are listed in Table 6.1 of Reference 8. So, the element values for the filters with the same order and ripple level as its dual

TABLE 6.1 Element Values for Bandpass Filters for
$N = 2$ and $L_{Ar} = 0.1$ dB

FBW	θ_c	$z_1 = z_2$	$z_{1,2}$
1.56	25°	0.154	1.135
1.33	30°	0.220	1.116
1.22	35°	0.308	1.090
1.11	40°	0.420	1.054
1.00	45°	0.569	1.008
0.89	50°	0.762	0.949
0.78	55°	1.015	0.876
0.67	60°	1.355	0.788
0.56	65°	1.822	0.686
0.44	70°	2.500	0.569
0.33	75°	3.586	0.439
0.22	80°	5.668	0.299
0.11	85°	11.698	0.151

TABLE 6.2 Element Values for Bandpass Filters for
$N = 3$ and $L_{Ar} = 0.1$ dB

FBW	θ_c	z_1	z_2	$z_{1,2}$
1.44	25°	0.197	0.182	1.121
1.33	30°	0.286	0.307	1.092
1.22	35°	0.401	0.483	1.054
1.11	40°	0.547	0.719	1.005
1.00	45°	0.730	1.028	0.945
0.89	50°	0.960	1.426	0.874
0.78	55°	1.253	1.941	0.792
0.67	60°	1.635	2.614	0.700
0.56	65°	2.150	3.525	0.599
0.44	70°	2.893	4.832	0.490
0.33	75°	4.080	6.910	0.374
0.22	80°	6.367	10.886	0.252
0.11	85°	13.039	22.415	0.127

network in Figure 6.2 are exactly unchanged. The difference is that normalized admittances in the former case need to be swapped over with normalized impedances in the later case. Both of them share the same transfer function.

On the other hand, the element values of the corresponding lowpass filters with series short-ended or shunt open-ended stubs can also be found from Tables 6.1 to 6.4, where θ_c needs to be replaced by $(\pi/2 - \theta_c)$ from Tables 6.1 to 6.4. Since the element values used in the above

TABLE 6.3   Element Values for Bandpass Filters for
$N = 2$ and $L_{Ar} = 1$ dB

FBW	θ_c	$z_1 = z_2$	$z_{1,2}$
1.56	25°	0.453	1.552
1.33	30°	0.624	1.508
1.22	35°	0.835	1.452
1.11	40°	1.096	1.383
1.00	45°	1.418	1.299
0.89	50°	1.820	1.202
0.78	55°	2.329	1.089
0.67	60°	2.991	0.964
0.56	65°	3.887	0.825
0.44	70°	5.181	0.675
0.33	75°	7.258	0.515
0.22	80°	11.276	0.348
0.11	85°	23.028	0.175

TABLE 6.4   Element Values for Bandpass Filters for
$N = 3$ and $L_{Ar} = 1$ dB

FBW	θ_c	z_1	z_2	$z_{1,2}$
1.44	25°	0.517	0.411	1.533
1.33	30°	0.712	0.647	1.480
1.22	35°	0.950	0.955	1.414
1.11	40°	1.238	1.348	1.335
1.00	45°	1.589	1.840	1.243
0.89	50°	2.020	2.457	1.140
0.78	55°	2.559	3.238	1.025
0.67	60°	3.255	4.247	0.899
0.56	65°	4.192	5.606	0.764
0.44	70°	5.546	7.557	0.622
0.33	75°	7.720	10.669	0.472
0.22	80°	11.940	16.658	0.318
0.11	85°	24.3170	34.118	0.160

synthesis procedure are normalized impedances with respect to the terminating impedance Z_0, the associated or real impedances for the stubs and lines are

$$Z_i = Z_0 z_i$$
$$Z_{i,i+1} = Z_0 z_{i,i+1}.$$

(6.29)

Example 6.1   Series stub UWB filter

This section describes the synthesis design and practical implementation of an UWB filter with series stubs. The specifications are given as the passband ripple level $L_{Ar} = 0.1$ dB and $\theta_c = 40°$ at cutoff frequency of 3 GHz, while the filter stage is readily chosen as $N = 3$. From Table 6.1, the element values under these specifications can be read out as

$$z_1 = z_3 = 0.54659 \quad z_2 = 0.71896 \quad z_{1,2} = z_{2,3} = 1.00474.$$

The filter network is terminated by $Z_0 = 50 \ \Omega$ at two external ports as usual, so the characteristic impedances of three series stubs and two connecting lines between adjacent stubs can be found as

$$Z_1 = Z_3 = 27.3 \ \Omega \quad Z_2 = 36 \ \Omega \quad Z_{1,2} = Z_{2,3} = 50.2 \ \Omega.$$

After these impedances are determined on an ideal transmission line network, one particular transmission line needs to be chosen in construction of such a filter network. In this aspect, a hybrid structure, composed of CPW and slotline, is chosen to design such a filter, and Figure 6.5 shows its layout. The feeding lines and connecting lines are implemented on CPW, while the series stubs are realized by open-ended slotlines. This hybrid structure is used here primarily due to its capability in easy realization of two functionalities in the implementation, that is, series installation and open termination of the transmission line stubs. As shown in Figure 6.3, a pair of slotline stubs in parallel is used to implement one series stub so as to keep the CPW portion symmetrically adopted. Based on the outcome of the above synthesis approach, the characteristic impedances of the stubs (Z_{o1}, Z_{o2}, and Z_{o3}) and connecting lines ($Z_{c1,2}$ and $Z_{c2,3}$) can be found as

$$Z_{o1} = Z_{o3} = 54.6 \ \Omega \quad Z_{o2} = 72 \ \Omega \quad Z_{c1,2} = Z_{c2,3} = 50.2 \ \Omega.$$

Meanwhile, the electrical lengths are found as $\theta_c = 40°$ for the stubs and $2\theta_c = 80°$ for the connecting lines. In practice, the filter is designed and implemented on Roger's RT/Duriod 6010 with the relative permittivity of $\varepsilon_r = 10.8$ and the thickness of $h = 0.635$ mm. The initial dimensions of the filter can be simply calculated using the well-known closed formula in Reference 9. In order to further take into account the effects of various discontinuities and frequency dispersion, optimization or tuning procedure has to be executed based on full-wave simulation software in the finals. Figure 6.6a shows a discontinuity at CPW and slotline junction. If the CPW line and slotline stubs are directly connected to each other, the frequency response, or, more specifically, the return loss, is not symmetrical as shown in Figure 6.6b. This will affect the balanced frequency response especially in the passband. Hence, a compensation technique has to be proposed, where the strip width of the CPW line is slightly narrowed around the junction part. From Figure 6.6b, we can observe that the transmission pole is moving to a low value as the strip width is reduced. From the circuitry point of view, the narrowed strip conductor in CPW provides an inductive effect so as to compensate for a parasitic capacitive effect occurring around the junction. The final filter dimension is tabulated in Table 6.5.

Figure 6.7a shows the photograph of the fabricated bandpass filter with series open-ended stubs. The measured S-parameters are plotted with the results obtained from simulation and transmission line network in Figure 6.7b. As one figure of merit for a wideband bandpass filter, three sets of group delay are also depicted in Figure 6.7c. The three sets of frequency responses agree well with each other, where five transmission poles are observed. However, some discrepancies are noticed, especially in the frequency range from DC to 2.4 GHz. In fact, the open-circuited slotline stubs should be considered as a coupling gap at low frequencies, thus its effect cannot be accurately included in the equivalent transmission-line model shown in Figure 6.3. It is the main reason why the measured and simulated $|S_{21}|$ fall much more slowly than that based on the transmission-line theory as the frequency is decreased to DC.

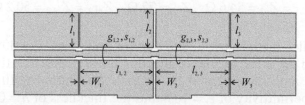

Figure 6.5. Layout of the UWB filter with series stubs ($N = 2$, $L_{Ar} = 0.1$ dB).

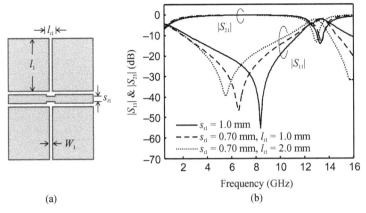

(a)   (b)

Figure 6.6. Compensation of cross-junction on slotline and CPW. (a) Layout. (b) Frequency responses.

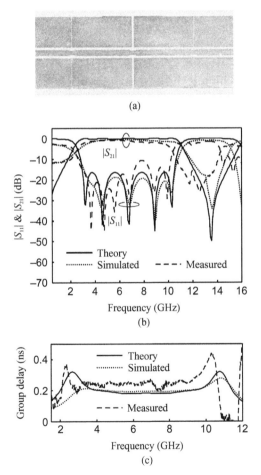

Figure 6.7. An UWB filter with hybrid CPW and slotline structures. (a) Photograph of the fabricated filter. (b) Predicted and measured S-parameters. (c) Predicted and measured group delays.

TABLE 6.5   Dimension of the Bandpass Filter Design in Example 6.1

$l_1 = l_3$ (mm)	l_2 (mm)	$l_{1,2} = l_{2,3}$ (mm)	$W_1 = W_3$ (mm)
5.10	5.20	10.18	0.11
W_2 (mm)	$g_{1,2} = g_{2,3}$ (mm)	$s_{1,2} = s_{2,3}$ (mm)	$l_{t1} = l_{t3}$ (mm)
0.33	0.38	1.0	1.0
l_{t2} (mm)	$s_{t1} = s_{t3}$	s_{t2}	
1.0	0.74	0.70	

Example 6.2   Series stub lowpass filter

As we mentioned early, this synthesis design method is also workable for the lowpass filter design with series short-ended stubs cascading through uniform transmission line sections. The element values of such a network, which is the antiparty of the highpass network shown in Figure 6.3, have already been listed in Tables 6.1–6.4. Therefore, it is not necessary to repeat the whole procedure. We only need to replace θ_c by $(\pi/2 - \theta_c)$ when the element values are read out from the table. Now, let us consider the design of a three-stage ($N = 3$) lowpass filter with $\theta_c = 45°$ at cutoff frequency $f_c = 3$ GHz and 0.1 dB ripple level. First, we can find the element values by looking up the row with $\theta_c = 90° - 45° = 45°$ in Table 6.2, such that

$$z_1 = z_3 = 0.72969 \quad z_2 = 1.02772 \quad z_{1,2} = z_{2,3} = 0.94472.$$

So, the characteristic impedances can be scaled by the terminal impedance of 50 Ω. To implement this lowpass filter, the series short-ended stubs are formed on slotline and the connecting lines are constructed on CPW. However, this filter can be more simply implemented on hybrid microstrip line and slotline structure as shown in Figure 6.8. The microstrip line and slotline are formed on the top and bottom metal layers, respectively, and the coupling is through microstrip-to-slotline transition. The equivalent circuit of this transition is modeled as a transformer with ratio n, which is a function of the strip and slot widths as well as the substrate thickness. Here, we assume $n = 1$. Again, one series stub is implemented by two slotline stubs in parallel. Thus, the characteristic impedances for the stubs and lines are

$$Z_{s1} = Z_{s3} = 73, \quad Z_{s2} = 102.8, \quad Z_{c1,2} = Z_{c2,3} = 74.2.$$

In this implementation, the substrate RT 6010, with a relative permittivity of 10.8 and thickness of 0.635 mm, is used. To compensate for the microstrip-to-slotline transition junction effect and increase the transformer ratio, the microstrip line on the top layer is narrowed at the transition portion. The whole lowpass filter layout is then optimally designed based on a full-wave electromagnetic simulator. Figure 6.9a shows the photographs of the fabricated lowpass filter. The simulated, measured and theoretical frequency responses are in good agreement with each other, as illustrated in Figure 6.9b.

6.4 TRANSMISSION LINE NETWORK WITH HYBRID SERIES AND SHUNT STUBS

In Section 6.3, we have introduced the UWB filters with cascaded series open-ended or shunt short-ended stubs through uniform transmission line sections with different lengths. As is well known, both the series open stub and shunt short stub serve as a highpass element. How about a network consisting of a mixed of both types of stubs instead of purely one type? In this section, we will study a transmission line network composed of both series open-circuited and shunt short-circuited stubs. A general transmission line model with composite stubs is illustrated in Figure 6.10. The network can be considered as three parts. Two identical sets of hybrid stubs structures are connected through a nonuniform transmission line with i sections, and i can be either odd in Figure

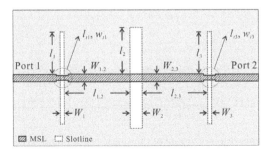

Figure 6.8. Schematic layout of a lowpass prototype filter ($N = 3$) implemented with hybrid microstrip line and slotline structures. Dimensions: $l_1 = l_3 = 5.91, l_2 = 6.50, l_{1,2} = l_{2,3} = 8.89$, $W_{1,2} = W_{2,3} = 0.33$, $W_1 = W_3 = 0.61$, $W_2 = 1.0$, $l_{t1} = l_{t3} = 1.0$, $w_{t1} = w_{t3} = 0.38$. All units are in mm.

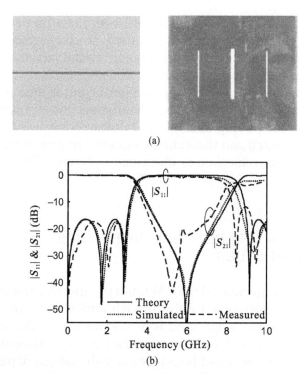

(a)

(b)

Figure 6.9. Lowpass filter ($N = 3$) using hybrid microstrip line and slotline structures. (a) Photograph of the fabricated filter. (b) Predicted and measured S-parameters.

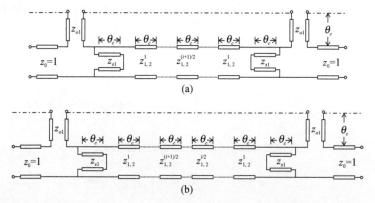

Figure 6.10. Transmission line model of the single-stage bandpass filter based on hybrid stubs and an i-section nonuniform connecting line. (a) i is an odd integer (1, 3, 5...). (b) i is an even integer (0, 2, 4, 6 ...).

6.10a or even in Figure 6.10b. The hybrid stubs are composed of a series open stub and a shunt short stub. All the series and shunt stubs have the same electrical length, θ_c, which is the electrical length at lower cutoff frequency, f_c. The normalized impedances for the series and shunt stubs are indicated as z_{om} and z_{sm}, where m indicates the m-th set of composite stubs. Meanwhile, each section of the middle connecting line is exactly chosen as θ_c, and the normalized impedances of the first section to the one at the centre are named from $z_{m,m+1}^1$ to $z_{m,m+1}^{(i+1)/2}$ for the odd-integer i (1, 3, 5, ...) and to $z_{m,m+1}^{i/2}$ for the even-integer i (0, 2, 4, 6, ...). The subscript indicates the connecting line between the m-th and $(m + 1)$-th sets of stubs, and the superscript is the section number. The network is symmetrical, while the input and output are both terminated by 50 Ω, that is, $z_0 = 1$.

Following the synthesis approach described above, the overall $ABCD$ matrix of the entire network shown in Figure 6.10 can be obtained by multiplying the individual $ABCD$ matrix of each unit section in consequence. Using Equation (6.13), the squared magnitude of transmission coefficient S_{21} can be expressed as

$$|S_{21}|^2 = \frac{1}{1 + \left| j\dfrac{P_{i+3}(\cos\theta)}{\sin^3\theta} \right|^2}, \tag{6.30}$$

where P_{i+3} is a polynomial function with highest order of $(i + 3)$.

It is remarked that the highest order indicates the number of transmission poles in the passband. As can be seen in Equation (6.30), the highest order of the sine function in the denominator is 3. As such, transmission zero emerges at multiple integer times of π with the third-order transmission zeros at highest. On the other hand, the transfer function in Equation (6.9) results in a Chebyshev equal-ripple response with $(n + q)$ in-band transmission poles and generates q^{th}-order transmission zeros. In order to make these two equations equal, we can get

$$\begin{aligned} n &= i \\ q &= 3. \end{aligned} \tag{6.31}$$

In this way, there will be $(i + 4)/2$ or $(i + 5)/2$ sets of equations derived depending on whether i is even or odd. By specifying the required ripple constant ε and cutoff phase θ_c or FBW, the unknown element values in the network can be determined as well addressed above.

A bandpass filter network with $i = 1$ is illustrated in Figure 6.11. There is only one connecting line section between two hybrid stubs.

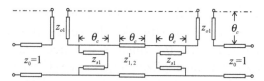

Figure 6.11. Transmission line model of bandpass filter with hybrid stubs and $i = 1$.

The element values involved in this model are denoted as z_{o1}, z_{s1}, and $z^1_{1,2}$. As a consequence, the characteristic function F can be derived from the transmission-line network as

$$F = j\left(k_1 \frac{\cos^4 \theta}{\sin^3 \theta} + k_2 \frac{\cos^2 \theta}{\sin^3 \theta} + k_3 \frac{1}{\sin^3 \theta} \right), \tag{6.32}$$

and

$$k_1 = \frac{z_{o1}^2 z^1_{1,2} - z^1_{1,2}}{2z_{s1}^2} + \frac{z_{o1}^2 + z_{o1}z^1_{1,2} - 1}{z_{s1}} + \frac{z_{o1}^2 - 1}{2z^1_{1,2}} + z_{o1} + \frac{z^1_{1,2}}{2} \tag{6.33a}$$

$$k_2 = -\frac{z_{o1}^2 - 2}{2z^1_{1,2}} - \frac{z_{o1}z^1_{1,2} - 1}{z_{s1}} - z_{o1} - z^1_{1,2} + \frac{z^1_{1,2}}{2z_{s1}^2} \tag{6.33b}$$

$$k_3 = \frac{z^1_{1,2}}{2} - \frac{1}{2z^1_{1,2}}. \tag{6.33c}$$

From Equation (6.31), the n and q in Equation (6.9) should be chosen as 1 and 3. Hence,

$$\cos(\phi + 3\xi) = T_1(x)T_3(y) - V_1(x)V_3(y), \tag{6.34}$$

and we have

$$T_1(x) = x \tag{6.35a}$$

$$V_1(x) = -\frac{1}{\sqrt{1 - x^2}}(x^2 - 1) \tag{6.35b}$$

$$T_3(y) = 4y^3 - 3y \tag{6.35c}$$

$$V_3(y) = -\frac{1}{\sqrt{1 - y^2}}(4y^4 - 5y^2 + 1). \tag{6.35d}$$

By substituting Equation (6.35) into Equation (6.34), and manipulating the equation, we have

$$\varepsilon\cos(\phi+3\xi) = \varepsilon\left(4\alpha(\alpha^2-1)^{\frac{3}{2}} + 4\alpha^4 + 3\alpha\sqrt{\alpha^2-1} - 3\alpha^2\right)\frac{\cos^4\theta}{\sin^3\theta}$$

$$+ \varepsilon\left(-3\alpha\sqrt{\alpha^2-1} - 5\alpha^2 + 3\right)\frac{\cos^2\theta}{\sin^3\theta} + \frac{\varepsilon}{\sin^3\theta}. \qquad (6.36)$$

It should be reminded that Equations (6.32) and (6.36) have the exact polynomial format. To get the same magnitude of two equations, we need to equalize all the coefficients for the terms with different orders. So, the following equations can be derived as

$$\varepsilon\left(4\alpha(\alpha^2-1)^{\frac{3}{2}} + 4\alpha^4 + 3\alpha\sqrt{\alpha^2-1} - 3\alpha^2\right) = \frac{z_{o1}^2 z_{1,2}^1 - z_{1,2}^1}{2z_{s1}^2} + \frac{z_{o1}^2 + z_{o1}z_{1,2}^1 - 1}{z_{s1}}$$

$$+ \frac{z_{o1}^2 - 1}{2z_{1,2}^1} + z_{o1} + \frac{z_{1,2}^1}{2}$$

$$\qquad (6.37a)$$

$$\varepsilon\left(-3\alpha\sqrt{\alpha^2-1} - 5\alpha^2 + 3\right) = -\frac{z_{o1}^2 - 2}{2z_{1,2}^1} - \frac{z_{o1}z_{1,2}^1 - 1}{z_{s1}} - z_{o1} - z_{1,2}^1 + \frac{z_{1,2}^1}{2z_{s1}^2} \quad (6.37b)$$

$$\varepsilon = \frac{z_{1,2}^1}{2} - \frac{1}{2z_{1,2}^1}. \qquad (6.37c)$$

In this way, the normalized characteristic impedances, z_{o1}, z_{s1}, and $z^1{}_{1,2}$, for the series open-ended stub, shunt short-ended stub, and the middle connecting line can be explicitly determined once ε and θ_c or *FBW* are specified. As indicated in Equation (6.37c), the normalized characteristic impedance of the connecting line with a length of θ_c only relies on the specified ripple constant ε. Moreover, the smaller ripple level in the passband requires smaller characteristic impedance of the middle line. The element values for the hybrid stub-based UWB filters in Figure 6.10 with $i = 1$ and 2, ripple level $L_{Ar} = 0.1$ dB, and $\theta_c = 20°$–$65°$, are tabulated in Tables 6.6 and 6.7.

The length of the nonuniform connecting line between two sets of hybrid stubs dominates the number of transmission poles in the passband. Figure 6.12a compares the $|S_{21}|$ of the prototype filter with different *i*, L_{Ar}, and θ_c are 0.1 dB and 45° as predetermined. It is no doubt that the number of transmission poles in the passband equals to $(i + 3)$. The extra transmission poles are introduced by an additional section in the connecting line. The filter with more transmission poles in the passband offers sharper rejection skirts under the same ripple level and cutoff phase. On the other hand, if a specified roll-off rate at the

TABLE 6.6   Element Values for Bandpass Filters for $n = 1$, $q = 3$ and $L_{Ar} = 0.1$ dB

0.1 dB Ripple				
FBW	θ_c	z_{o1}	z_{s1}	$z_{1,2}$
1.56	20	0.386	4.054	1.164
1.44	25	0.497	3.033	1.164
1.33	30	0.619	2.350	1.164
1.22	35	0.755	1.859	1.164
1.11	40	0.901	1.486	1.164
1.00	45	1.088	1.194	1.164
0.89	50	1.301	0.959	1.164
0.78	55	1.565	0.765	1.164
0.67	60	1.904	0.602	1.164
0.56	65	2.363	0.464	1.164

TABLE 6.7   Element Values for Bandpass Filters for $n = 2$, $q = 3$ and $L_{Ar} = 0.1$ dB

0.1 dB Ripple				
FBW	θ_c	z_{o1}	z_{s1}	$z_{1,2}$
1.56	20	0.394	3.491	1.210
1.44	25	0.510	2.539	1.240
1.33	30	0.637	1.918	1.282
1.22	35	0.778	1.486	1.337
1.11	40	0.938	1.169	1.410
1.00	45	1.124	0.930	1.506
0.89	50	1.345	0.743	1.634
0.78	55	1.619	0.593	1.807
0.67	60	1.969	0.470	2.049
0.56	65	2.445	0.367	2.398

passband edges is required, the filter with more poles can widen the bandwidth of the core passband at no cost of extra resonators. As i increases from 1 to 3, two extra poles are introduced into the passband. As a result, the FBW with the same selectivity is greatly extended from 103.9 to 142.0%. Figure 6.12b shows the frequency responses of two bandpass filters with the same rejection level of 20.0 dB but different bandwidths.

It is very interesting to give a comparative study on the hybrid stub filter with the one using pure series or shunt stubs. Observing the polynomial functions in Equation (6.32), one can understand that the

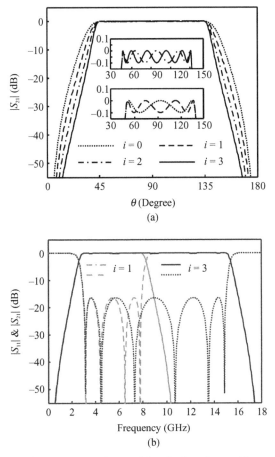

Figure 6.12. Comparison among the multi-pole bandpass filters with different i ($L_{Ar} = 0.1$ dB). (a) Improved selectivity with the same $\theta_c = 45°$. (b) Extendable bandwidth with the same rejection skirts.

number of in-band transmission poles equals to the highest order of $\cos\theta$ in the numerator of characteristic function F. The highest order of $\sin\theta$ in denominator is $q = 3$, and it represents the existence of the third-order transmission zeros in the lower and upper ends of the passband, that is, $\theta = 0°$ and $180°$. It needs to be highlighted that there exist only first-order transmission zeros in the high-pass prototypes with purely shunt/series stubs in Equation (6.16). Thus, the out-of-band roll-off or attenuation skirt gets significantly enhanced, especially for the multi-stage filter prototype with multiplying attenuation that will be introduced later. Figure 6.13 depicts the performances of these two

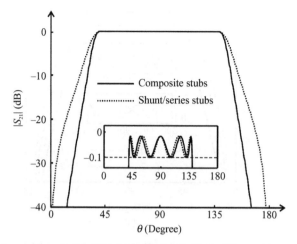

Figure 6.13. Comparison in $|S_{21}|$ between the five-pole bandpass filter prototypes with purely shunt stubs or series stubs and hybrid stubs.

types of bandpass filters under the same specifications and existence of five transmission poles in the constituted passband. It is quantitatively confirmed that the proposed filter prototype with composite series/ shunt stubs has much higher attenuation skirt in the lower stopband of $0 < \theta < \theta_c$ and the upper stopband band of $(180° - \theta_c) < \theta < 180°$ than those with either series or shunt stubs.

Example 6.3   Hybrid stub UWB filter

To illustrate the design procedure of this type of UWB filter with composite stubs, let us consider a five pole ($i = 2$) Chebyshev bandpass filter with 0.1 dB passband ripple and $FBW = 111\%$ at the center frequency $f_0 = 6.75$ GHz. To start with this design, we need to find out the element values for this filter with $i = 2$. Using Equation (6.3), the cutoff phase θ_c can be calculated as

$$\theta_c = 90° - 45° \times FBW = 40°.$$

Looking at Table 6.7, the element values are found as

$$z_{o1} = 0.93779 \quad z_{s1} = 1.16933 \quad z^1_{1,2} = 1.40952.$$

As two ports are terminated by 50 Ω, the characteristic impedances for the stubs and lines are

$$Z_{o1} = 46.89 \quad Z_{s1} = 58.47 \quad Z_{1,2}^1 = 70.48.$$

Since the center frequency is 6.75 GHz, the cutoff frequency f_c is

$$f_c = \frac{\theta_c}{90°} f_0.$$

Without loss of generality, this unique type of transmission line network is formed on hybrid microstrip and slotline structure. As shown in the photograph of the constituted filter in Figure 6.15, the series open-ended stub and the shunt short-ended stub are realized with virtue to a microstrip-to-slotline transition. To simplify our analysis, the ratio of the transformer is assumed to be 1. The connecting line between two sets of stubs is formed as a slotline on ground plane. The above substrate with relative permittivity of $\varepsilon_r = 10.8$ and thickness of $h = 0.635$ mm is still used in this design. After the initial dimensions of the microstrip line and slotline are calculated by the close formula [9], a full-wave simulator is employed to determine its final filter layout with the detailed dimensions denoted in Figure 6.14. Figure 6.15a is the photographs of the front and back sides of the fabricated filter. The frequency responses and group delay from the ideal transmission line model in Figure 6.12 are depicted with the simulated and measured ones in Figure 6.15b,c. The three sets of curves show a good agreement among themselves.

Now, let us compare the general UWB filter prototype shown in Figure 6.10 with the basic UWB bandpass filter discussed in

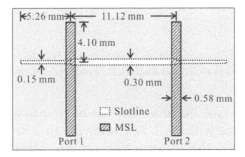

Figure 6.14. Layout of the UWB bandpass filters with hybrid stubs ($n = 2$, $q = 3$, and $L_{Ar} = 0.1$ dB).

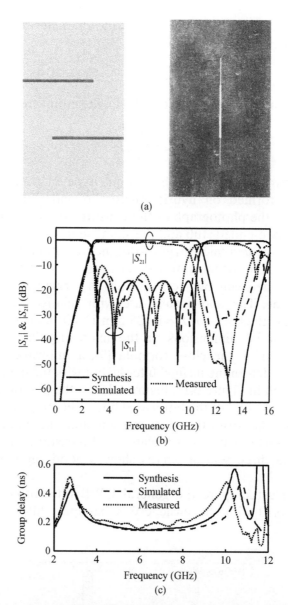

Figure 6.15. (a) Photographs of the fabricated UWB filter with hybrid stubs. (b) Predicted and measured S-parameters. (c) Predicted and measured group delays.

Section 5.4.2. In fact, the latter one can be treated as a specified case of the former one with $i = 2$. Following the discussion in Chapter 5, the key idea in the construction of an MMR-based filter is to consider the overall slotline section as a triple-mode resonator with uniform and nonuniform configurations. In particular, the non-

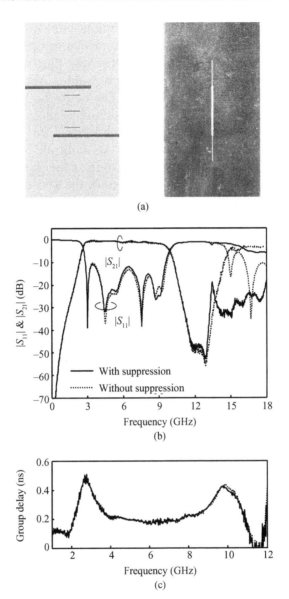

Figure 6.16. Comparison between measured results of two five-pole bandpass filters with and without periodical loading of three strips. (a) Photographs. (b) S-parameters. (c) Group delays.

uniform section with different characteristic impedances provides more flexibility and capability in the formulation of a desired wide passband.

With proper adjustment of the impedance ratio, the first three resonances occur at equally spaced frequencies within the specified passband. Together with the tight coupling behavior of microstrip-to-slotline transition, two additional transmission poles can be produced so as to realize a five-pole ultra-wide passband with varied bandwidths. By observing the element values obtained from the above synthesis design in Table 6.7, it can be found that as the *FBW* increases, the normalized characteristic impedances become higher for the two shunt short-ended stubs and lower for the middle line. This phenomenon exactly matches the multiple-mode-resonance theory as we discussed in Chapter 5. On the other hand, in the transmission line model shown in Figure 6.10, the short-/open-circuited stubs have the electrical length θ_c at lower cutoff frequency f_c, that is, equal to 90° at the center frequency of the passband. Thus, this approach has well explained and verified the multiple-pole frequency response of the MMR-based UWB filter from the synthesis design point of view. Therefore, instead of time-consuming cut-and-try method via full-wave simulators, this synthesis method allows us to efficiently design these wideband filters using a set of closed-form design formula under the prior specifications.

To further improve the upper stopband performance, a slow-wave line with periodical interruption along the longitudinal direction is utilized to replace the middle connecting line shown in Figure 6.10. For a demonstration in design procedure, let us reconsider the filter in Example 6.3. The middle line is periodically interrupted by transverse strips on the opposite interface of a slotline resonator, as shown in Figure 6.16. In this aspect, the strip-interrupted periodic slotline actually act as one effective transmission line at low frequencies covering the concerned frequency band. Figure 6.16 shows the frequency responses of the two filters, with and without attaching the strips. Obviously, the in-band frequency responses are almost unchanged, but the upper stopband is significantly pushed from 14.0 to 18.0 GHz so as to make up a good upper stopband until 18.0 GHz, with an attenuation of higher than 18.82 dB.

The hybrid stub UWB filter can also be constructed with multistage configuration. Figure 6.17 shows a three-stage bandpass filter prototype with $i = 1$. This filter is composed of four sets of composite stubs that are cascaded through three sections of transmission line. All the series stubs, shunt stubs, and connecting lines have a fixed electrical length, that is, θ_c, at the specified lower cutoff frequency f_c. The normalized characteristic impedances of the first and second sets of hybrid stubs

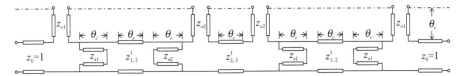

Figure 6.17. Transmission line model of a three-stage ($N = 3$) bandpass filter prototype with hybrid stubs ($i = 1$).

TABLE 6.8   Design Parameters for the Three-Stage Hybrid UWB Filter

Stub Line Impedance	Connecting Line Impedance
$Z_{o1} = 86.80\ \Omega$	$Z^1_{1,2} = 75.46\ \Omega$
$Z_{s1} = 74.20\ \Omega$	$Z^1_{2,3} = 70\ \Omega$
$Z_{o2} = 94.43\ \Omega$	$Z_0 = 70\ \Omega$
$Z_{s2} = 50.40\ \Omega$	

are denoted as z_{o1}, z_{s1} and z_{o2}, z_{s2}, respectively, while the ones for the connecting lines between the first set and second set of stubs, the second and the third set of stubs, are $z^1_{1,2}$ and $z^1_{2,3}$. The *ABCD* matrix of the whole network can be obtained by multiplying the *ABCD* submatrices of all the sections in sequence. Therefore, the characteristic function *F* can be derived as

$$F = j\left(k_1 \frac{\cos^8 \theta}{\sin^5 \theta} + k_2 \frac{\cos^6 \theta}{\sin^5 \theta} + k_3 \frac{\cos^4 \theta}{\sin^5 \theta} + k_4 \frac{\cos^2 \theta}{\sin^5 \theta} + k_5 \frac{1}{\sin^5 \theta} \right), \quad (6.38)$$

where k_1, k_2, k_3, k_4, and k_5 are the coefficients of the polynomial, and they are functions of the six element variables

$$k_1, k_2, k_3, k_4, k_5 = f\left(z_{o1}, z_{s1}, z^1_{1,2}, z_{o2}, z_{s2}, z^1_{2,3}\right). \quad (6.39)$$

This characteristic function *F* has a similar format as that discussed before, except that the highest order in numerator is 8 while that in denominator is 5. Comparing Equation (6.9) with Equation (6.38), the values for *n* and *q* are selected as 3 and 5. Five equations can be derived in terms of six variables. In this aspect, the element values are determined in such a way that the frequency response derived from the network is approaching the targeted transfer function. The calculated design parameters are listed in Table 6.8. It is remarked that the transmission zeros appeared in multiple integer times of π are

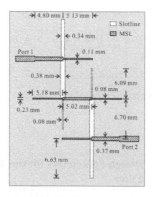

Figure 6.18. Layout of the three-stage UWB bandpass filter shown in Figure 6.18.

fifth-order transmission zeros. Thus, if both single stage and three-stage bandpass filters are designed under the same specifications, that is, ripple level, lower cutoff phase, and same number of transmission poles, the out-of-band rejection skirt of the three-stage filter definitely becomes sharper at both lower and higher cut-off edges of the passband.

To implement this type of multistage hybrid stub filter, microstrip combined with slotline is selected. The series open stubs are all constructed by the open-circuited microstrip lines, while the shunt short stubs are all formed by the slotline on ground plane. The uniform lines between two sets of stubs can be either microstrip line or slotline. Due to physical constraint in a large value (e.g., >100 Ω) of characteristic impedances using this substrate filter layout, the terminal impedances are chosen as 70 Ω. To connect this filter with two 50-Ω coaxial cables in measurement, two quarter-wavelength transformers are additionally installed along the main feed lines, as shown in Figure 6.18. The same substrate with relative dielectric constant of 10.8 and substrate thickness of 0.635 mm is used in this design as well.

To accurately take into account the unexpected effects due to discontinuities, frequency dispersion and radiation, a full-wave simulator is used to carry out the final-stage design of the filter performance, leading to derive the final filter dimensions, which are denoted in Figure 6.18. Figure 6.19 shows the front and back sides' photographs of a fabricated three-stage hybrid filter. The theoretical, simulated and measured filter responses are plotted together in Figure 6.19b,c for quantitative comparison.

From the MMR's point of view, this constituted filter is realized by three distinct resonators, that is, one microstrip line resonator and two

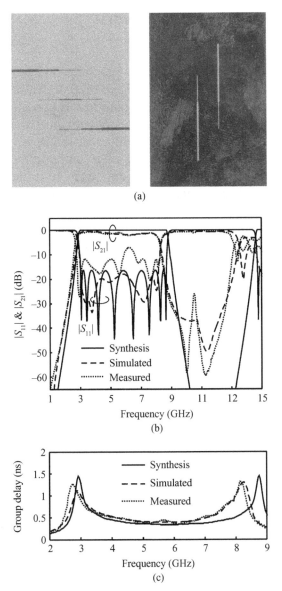

Figure 6.19. Three-stage UWB bandpass filter ($n = 3$, $q = 5$, and $L_{Ar} = 0.1$ dB). (a) Photographs. (b) Predicted and measured S-parameters. (c) Predicted and measured group delays.

slotline resonators, which are coupled via microstrip-to-slotline transitions. The first two resonant modes of these three resonators contribute to six transmission poles. In addition to the two poles from input/output coupling, this filter actually has eight poles in the passband.

6.5 MMR-BASED UWB FILTER WITH PARALLEL-COUPLED LINES

The initial MMR-based UWB filter in Figure 5.12 is now reanalyzed and redesigned using the above synthesis method. The exact equivalent circuit of one section of parallel-coupled line is two series open-ended stubs which are cascaded through a connecting line, as shown in Figure 6.20b [10]. The characteristic impedances of the series stubs and connecting line are indicated in terms of even- and odd-mode characteristic impedances of a standard coupled line. It is noticed that this equivalent circuit has a similar format as the transmission line network in Figure 6.4 except that the length of the connecting line is θ instead of 2θ. Again, all the characteristic impedances in the equivalent circuit are normalized by Z_0. Using the derivation procedure in Section 6.2, the characteristic function F of this network is derived as

$$
F = -j\left[\left(\frac{z_{0o}^2}{z_{0e} - z_{0o}} + z_{0o} + \frac{z_{0e} - z_{0o}}{4} - \frac{1}{z_{0e} - z_{0o}}\right)\frac{\cos^2\theta}{\sin\theta} \\
- \left(\frac{z_{0e} - z_{0o}}{4} - \frac{1}{z_{0e} - z_{0o}}\right)\frac{1}{\sin\theta}\right].
$$

(6.40)

where z_{0e} and z_{0o} are the normalized even- and odd-mode impedances of parallel-coupled line.

From Equation (6.40), the highest order of the characteristic function F is found as 2. Comparing Equation (6.40) with Equation (6.23), the highest order of polynomial is reduced by 1 as the central connecting line length is changed from 2θ to θ. Consequently, the number of transmission poles in the passband reduces to 2 from 3. It means that

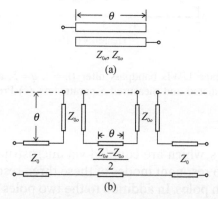

Figure 6.20. Parallel coupled line. (a) Schematic. (b) Equivalent circuit.

the number of poles is affected by the length of the connecting line as discussed in Section 6.4. Now, we know this equivalent circuit should generate two poles in the passband, which corresponds to the "over-match" condition in Equation (4.36). When the condition becomes "match," we have the following equation from Equations (4.35) and (4.36):

$$z_{0e} - z_{0o} = 2. \tag{6.41}$$

By substituting Equation (6.41) into Equation (6.40), the coefficient of the second term becomes zero. It implies that the two roots of equation $|F| = 0$ are the same as each other. Therefore, the two transmission poles in the passband merge so as to create only one pole in the "match" condition, as shown in Figure 4.22.

Let us take into account a MMR-based UWB filter with its first resonant mode occurring in its passband as an example to illustrate how this synthesis method can be applied to this type of UWB filters. Figure 6.21a shows the schematic layout of the MMR-based UWB filter. This filter is composed of a single transmission line resonator with parallel-coupled line sections at two ports. In terms of its geometry, this filter is similar to a classical single-pole filter that consists of two cascaded parallel-coupled lines. However, the filter presented here can generate three transmission poles rather than one pole in the desired passband.

Using this exact equivalent circuit shown in Figure 6.20b, the equivalent circuit of this MMR-based UWB filter can be alternatively depicted in Figure 6.21b. The lengths of the parallel-coupled line sections are all set as $\lambda_g/4$ at the center frequency of the passband, so they are actually

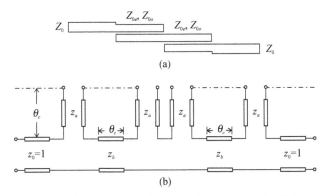

Figure 6.21. MMR-based UWB filter with a single resonator. (a) Schematic. (b) Equivalent circuit.

equivalent to θ_c at lower cutoff frequency, as shown in Figure 6.21b. The impedances of each section are normalized to port impedance, that is, $z_0 = 1$. For the convenience of derivation, the normalized impedances for the series stubs and connecting line are expressed as

$$z_a = Z_{0o} / Z_0 \tag{6.42a}$$

$$z_b = \frac{Z_{0e} - Z_{0o}}{2Z_0}. \tag{6.42b}$$

From the procedure described before, the characteristic function F of the network shown in Figure 6.21b can be derived:

$$F = j\left[-\left(3z_a - \frac{z_a}{z_b^2} + \frac{z_a^3}{z_b^2} - \frac{1}{z_b} + \frac{3z_a^2}{z_b} + z_b \right) \frac{\cos^3 \theta}{\sin \theta} + \left(z_a - \frac{z_a}{z_b^2} + z_b - \frac{1}{z_b} \right) \frac{\cos \theta}{\sin \theta} \right]. \tag{6.43}$$

It is interesting to see that Equation (6.43) has the same format of a polynomial function as Equation (6.23), and both have three transmission poles in the passband and first-order transmission zeros. Therefore, they should share the same targeted function (Eq. 6.27). By equalizing the magnitudes of these two functions, Equations (6.43) and (6.27), we can obtain

$$3z_a - \frac{z_a}{z_b^2} + \frac{z_a^3}{z_b^2} - \frac{1}{z_b} + \frac{3z_a^2}{z_b} + z_b = 2\varepsilon \left(\frac{1 + \sqrt{1 - \cos^2 \theta_c}}{\cos^3 \theta_c} \right) \tag{6.44a}$$

$$z_a - \frac{z_a}{z_b^2} + z_b - \frac{1}{z_b} = \varepsilon \left(\frac{2 + \sqrt{1 - \cos^2 \theta_c}}{\cos \theta_c} \right). \tag{6.44b}$$

By solving Equation (6.44) and substituting z_a and z_b into Equation (6.42), even- and odd-mode impedances of the parallel-coupled line in Figure 6.21a can be calculated. As discussed before, the proper stretch of the connecting line can bring out more poles in the passband. The schematic layout of the general MMR-based UWB filter is shown in Figure 6.22, and the element values for the filter with different m, ripple level L_{Ar}, and θ_c are listed in Tables 6.9–6.11.

As mentioned in Section 6.4, one major drawback of the synthesis design is that it cannot take account of the frequency dispersion that happens in any inhomogeneous transmission line. In fact, the impedance and propagation constant of all the transmission line sections

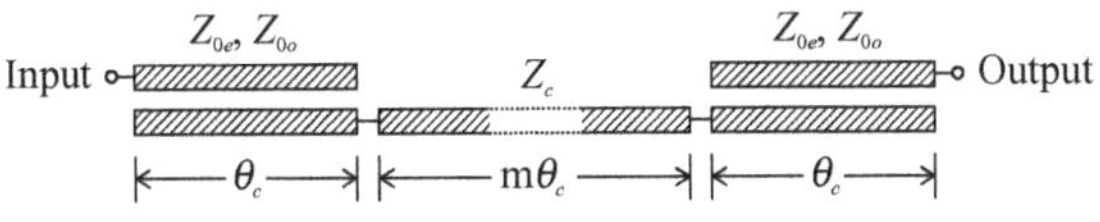

Figure 6.22. Ultra-wideband parallel-coupled line filter using single line resonator. ($m = 0$: three transmission poles; $m = 1$: four transmission poles; $m = 2$: five transmission poles.)

TABLE 6.9   Element Values for MMR-Based UWB Filters in Figure 6.22 with $m = 0$

θ_c	0.1 dB Ripple Level		0.2 dB Ripple Level		0.5 dB Ripple Level	
	Z_{0e}	Z_{0o}	Z_{0e}	Z_{0o}	Z_{0e}	Z_{0o}
5	117.528	0.394	125.607	0.558	143.363	0.893
10	118.879	0.934	127.503	1.318	146.343	2.095
15	120.520	1.666	129.778	2.339	149.863	3.690
20	122.512	2.654	132.505	3.703	154.008	5.787
25	124.935	3.982	135.781	5.515	158.894	8.528
30	127.900	5.766	139.730	7.917	164.674	12.097
35	131.555	8.166	144.524	11.100	171.563	16.742
40	136.106	11.404	150.403	15.330	179.865	22.805
45	141.851	15.800	157.713	20.983	190.020	30.768
50	149.231	21.826	166.976	28.618	202.704	41.347
55	158.945	30.213	179.014	39.094	218.983	55.650
60	172.151	42.159	195.207	53.831	240.651	75.512
65	190.948	59.803	218.047	75.361	270.951	104.214
70	219.540	87.388	252.539	108.725	316.397	148.303
75	267.759	134.783	310.388	165.660	392.215	223.034
80	365.143	231.554	426.743	281.343	544.117	374.143
85	659.436	525.470	777.422	631.610	1000.742	830.285

involved in this planar filter seem to vary tremendously across the ultra-wide bandwidth. Thus, the simulated and measured bandwidth of a UWB bandpass filter is usually smaller than that predicted from its ideal transmission line network. To circumvent this problem, a modified electrical length θ_m is introduced as

$$\theta_m = \theta_c \left(1 - \frac{FBW}{10} \right). \tag{6.45}$$

This modification is executed to compensate for bandwidth decrement, since θ_c is only accurate at low cutoff frequency of the passband. Let us take the MMR-based UWB filter in Figure 6.22 with $m = 1$ as an example. Figure 6.23 shows frequency responses of three sets of simulated results in comparison with the desired, classical θ_c, and modified

TABLE 6.10 Element Values for MMR-Based UWB Filters in Figure 6.22 with $m = 1$

θ_c	0.1 dB Ripple Level			0.2 dB Ripple Level			0.5 dB Ripple Level		
	Z_{0e}	Z_{0o}	Z_{12}	Z_{0e}	Z_{0o}	Z_{12}	Z_{0e}	Z_{0o}	Z_{12}
5	117.548	0.426	58.9140	125.632	0.603	63.014	143.394	0.961	72.014
10	118.942	1.088	59.653	127.575	1.527	64.045	146.420	2.404	73.623
15	120.620	2.078	60.351	129.877	2.885	65.008	149.929	4.475	75.101
20	122.607	3.513	60.915	132.563	4.816	65.783	153.937	7.340	76.285
25	124.944	5.538	61.235	135.676	7.481	66.245	158.500	11.189	77.031
30	127.700	8.327	61.202	139.300	11.069	66.283	163.735	16.237	77.227
35	130.993	12.085	60.725	143.581	15.802	65.817	169.840	22.740	76.810
40	135.016	17.066	59.7500	148.759	21.956	64.815	177.134	31.028	75.776
45	140.081	23.596	58.2755	155.207	29.898	63.296	186.109	41.560	74.170
50	146.683	32.138	56.3505	163.510	40.165	61.328	197.515	55.019	72.077
55	155.608	43.411	54.0650	174.589	53.602	59.006	212.528	72.497	69.605
60	168.162	58.625	51.5310	189.967	71.639	56.440	233.086	95.838	66.867
65	186.659	79.993	48.8644	212.353	96.891	53.739	262.654	128.413	63.969
70	215.685	112.003	46.1692	247.123	134.644	50.999	308.104	177.010	61.004
75	265.872	165.197	43.5297	306.754	197.297	48.295	385.421	257.543	58.048
80	369.018	271.304	41.0075	428.603	322.1425	45.686	542.486	417.833	55.157
85	683.975	589.120	38.6429	799.309	695.7750	43.210	1018.575	897.110	52.372

TABLE 6.11 Element Values for MMR-Based UWB Filters in Figure 6.22 with $m = 2$

θ_c	0.1 dB Ripple Level			0.2 dB Ripple Level			0.5 dB Ripple Level		
	Z_{0e}	Z_{0o}	Z_{12}	Z_{0e}	Z_{0o}	Z_{12}	Z_{0e}	Z_{0o}	Z_{12}
5	117.564	0.461	58.857	125.652	0.650	62.932	143.416	1.031	71.879
10	118.975	1.262	59.307	127.607	1.757	63.564	146.436	2.735	72.873
15	120.613	2.552	59.285	129.839	3.494	63.570	149.808	5.316	72.959
20	122.448	4.502	58.491	132.312	6.043	62.628	153.490	8.963	71.795
25	124.486	7.280	56.689	135.046	9.565	60.533	157.534	13.831	69.237
30	127.314	11.917	53.081	138.840	15.272	56.486	163.125	21.476	64.455
35	129.577	15.904	49.881	141.864	20.080	52.998	167.544	27.785	60.439
40	133.102	22.033	45.214	146.514	27.371	48.012	174.246	37.240	54.796
45	137.798	29.643	40.099	152.593	36.349	42.629	182.818	48.792	48.770
50	144.266	39.125	34.833	160.773	47.487	37.126	194.096	63.077	42.634
55	153.372	51.163	29.640	172.047	61.613	31.706	209.315	81.174	36.577
60	166.489	66.991	24.662	187.999	80.184	26.493	230.474	104.964	30.71370
65	186.014	88.906	19.966	211.413	105.899	21.541	261.101	137.896	25.094
70	216.669	121.569	15.562	247.798	144.211	16.857	308.202	186.926	19.725
75	269.428	175.914	11.421	309.961	207.895	12.414	388.082	268.326	14.579
80	377.158	284.787	7.491	436.286	335.315	8.164	549.616	430.955	9.615
85	704.137	612.455	3.708	818.641	718.330	4.048	1037.146	919.145	4.775

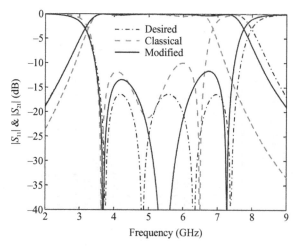

Figure 6.23. Frequency response of designed filters based on the desired, classical θ_c and proposed θ_m.

TABLE 6.12   Impedances of Coupled-Line Sections and Connecting Line Using the Classical and Modified Formulas

	Classical			Modified		
	Z_{0e} (Ω)	Z_{0o} (Ω)	Z_c (Ω)	Z_{0e} (Ω)	Z_{0o} (Ω)	Z_c (Ω)
$m = 1$	162.9	52.3	52.5	153.6	40.9	54.6
$m = 2$	143.4	38.0	35.4	137.8	29.6	40.1

θ_m. Herein, the desired 0.1 dB $FBW = 70.9\%$ is defined with reference to the low cutoff frequency at 3.6 GHz. The initial bandwidth, derived from θ_c, is less than 60%, and the bandwidth decrement becomes significant. Meanwhile, using the modified θ_m, the simulated bandwidth is about 69.4%, where the bandwidth decrement is only about 1.5%. Furthermore, the desired second and third poles are merged together. Although the deteriorated return loss is still acceptable in the majority of requirements, its responses can be recovered by slightly narrowing the widths of coupled line sections, which will be demonstrated later.

To verify the modified θ_m for more accurate prediction with a UWB bandwidth, that is, $FBW > 50\%$, two filters with $FBW = 70.9$ and 90.1% are designed on the RT/duroid 6010 substrate with the thickness 1.27 mm and the permittivity of 10.8. The required impedances of each section using the classical and modified formulas are tabulated in Tables 6.12 and 6.13.

It is noticed that the modified even-mode impedances of coupled-line sections are decreased by no more than 6%, while the odd-mode

TABLE 6.13 Impedances of Coupled-Line Sections and Connecting Line Using the Classical and Modified Formula

	Classical			Modified		
	Z_{0e} (Ω)	Z_{0o} (Ω)	Z_c (Ω)	Z_{0e} (Ω)	Z_{0o} (Ω)	Z_c (Ω)
$m = 1$	162.9	52.3	52.5	153.6	40.9	54.6
$m = 2$	143.4	38.0	35.4	137.8	29.6	40.1

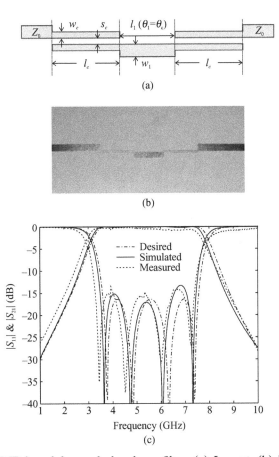

Figure 6.24. MMR-based four-pole bandpass filter. (a) Layout. (b) Photograph. (c) *S*-parameters.

impedance exhibits a significant change, that is, >20%, for both of the two cases. Meanwhile, the impedance of the connecting line becomes larger than the classical one. Figures 6.24 and 6.25 show the layouts and photographs of these designed MMR-based four-pole ($m = 1$) and five-pole ($m = 2$) bandpass filters, respectively. As shown in Figure 6.24, the

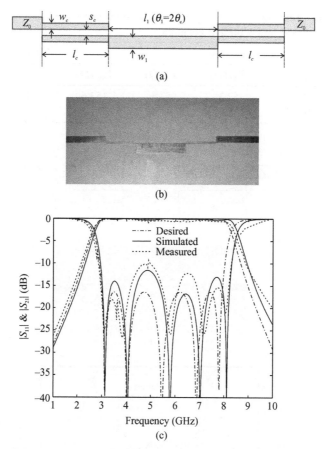

Figure 6.25. MMR-based five-pole bandpass filter. (a) Layout. (b) Photograph. (c) *S*-parameters.

simulated and measured four-pole filters achieve the fractional bandwidths of 71.6 and 74.5%, respectively.

On the other hand, fractional bandwidths of 92.1 and 92.4% are observed in the simulated and measured results of the designed filter in Figure 6.25. For both of the two filters, measured results show a good agreement with those obtained from simulation, and the measured bandwidth decrements are less than 3.6%, while the simulated ones are within 2.0%. The slight frequency shift between the simulated and measured results may be caused by the tolerance in etching fabrication for the narrow strip and slot widths. Again, the corrected number of transmission poles has been demonstrated in both cases as preferred, and it is realized by slightly narrowing the strip/slot widths of coupled-line sections.

6.6 SUMMARY

In this chapter, a direct synthesis approach for MMR-based UWB filters is presented. The insertion loss function is derived directly from the proposed transmission-line model and regulated to exhibit Chebyshev equal-ripple responses in the desired UWB passband. The general circuit model with series and shunt stubs is proposed and analyzed with the synthesis approach. By increasing the number of transmission-line sections between two stubs, one can increase the number of transmission poles in the passband, thus widening the concerned bandwidth. This approach provides the designers an alternative choice of constructing UWB bandpass filters.

REFERENCES

1 H. Ozaki and J. Ishii, "Synthesis of a class of strip-line filters," *IRE Trans. Circuit Theory* 5(2) (1958) 104–109.

2 H. Ozaki and J. Ishii, "Synthesis of transmission-line networks and the design of UHF filters," *IRE Trans. Circuit Theory* 2(4) (1955) 325–336.

3 H. J. Riblet, "The application of a new class of equal-ripple functions to some familiar transmission-line problems," *IEEE Trans. Microw. Theory Tech.* 12(4) (1964) 415–421.

4 R. Levy, "A new class of distributed prototype filters with applications to mixed lumped/distributed component design," *IEEE Trans. Microw. Theory Tech.* 18(12) (1970) 1064–1071.

5 M. C. Horton and R. J. Wenzel, "General theory and design of optimum quarter-wave TEM filters," *IEEE Trans Microw. Theory Tech.* 13(3) (1965) 316–327.

6 O. P. Gupta and R. J. Wenzel, "Design tables for a class of optimum microwave bandstop filters," *IEEE Trans. Microw. Theory and Tech.* 18(7) (1970) 402–404.

7 H. J. Carlin and W. Kohler, "Direct synthesis of band-pass transmission line structures," *IEEE Trans. Microw. Theory and Tech.* MTT-13 (1965) 283–297.

8 J.-S. Hong and M. J. Lancaster, *Microwave Filters for RF/Microwave Applications*, John Wiley & Sons, Inc, New York, 2001.

9 K. C. Gupta, R. Garg, I. Bahl, and P. Bhartia, *Microstrip Lines and Slotlines*, 2nd ed., Artech House, Norwood, MA, 1996.

10 G. Matthaei, L. Young, and E. M. T. Jones, *Microwave Filters, Impedance-Matching Network, and Coupled Structures*, Artech House, Dedham, MA, 1980.

CHAPTER 7

OTHER TYPES OF UWB FILTERS

7.1 INTRODUCTION

To meet the prompt needs in exploration of UWB technology in recent years, studies on UWB bandpass filters have increasingly become an attractive research topic since it is one of indispensable circuit blocks in the whole UWB system. Besides the MMR-based UWB filters discussed in the previous chapters, there are a few other types of UWB filters reported so far based on other design methodologies. In this chapter, these UWB filters will be generally classified and comparatively discussed in terms of their filtering performance, geometrical characteristic, operating mechanism, design procedure, and so on. As a result, the readers can have a global view on the current development of a variety of UWB bandpass filters.

7.2 UWB FILTERS WITH HIGHPASS AND LOWPASS FILTERS

The most straightforward way to design a UWB bandpass filter is to simply cascade a highpass filter with a lowpass filter [1–5]. The lower

Microwave Bandpass Filters for Wideband Communications, First Edition. Lei Zhu, Sheng Sun, Rui Li.
© 2012 John Wiley & Sons, Inc. Published 2012 by John Wiley & Sons, Inc.

and upper cutoff frequencies of the constituted UWB filter is predominantly determined by the cutoff frequencies of the highpass and lowpass filters, respectively. However, if the highpass and lowpass filters are directly connected, the overall circuitry area will be enlarged to a large extent. To circumvent this problem, a composite highpass–lowpass UWB filter was presented in Reference 1 by embedding a lowpass filter into a highpass filter. As shown in Figure 7.1a, the lowpass filter is made up of a stepped-impedance structure, while the highpass filter is a traditional prototype with shunt short-circuited stubs. These

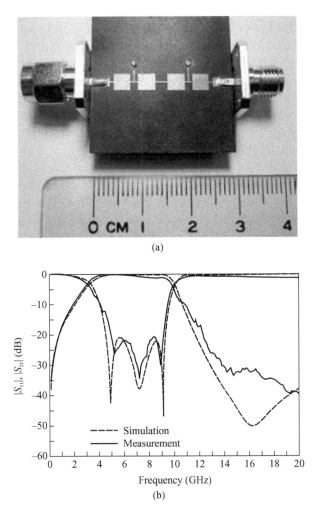

(a)

(b)

Figure 7.1. UWB filter composed of cascaded highpass and lowpass filters. (a) Photograph. (b) Simulated and measured frequency responses [1].

two individual filters are separately designed based on their respective cutoff frequencies. After the embedding procedure is properly executed, the final UWB bandpass filter can be constructed to achieve the expected UWB passband frequency response. In Reference 1, the proposed composite filter is implemented on a substrate with dielectric constant $\varepsilon_r = 2.2$ and thickness $h = 20$ mil. The measured results show an *FBW* about 108% at the center frequency of 6.5 GHz. The return loss within the passband is larger than 20 dB, and the upper stopband is up to 20 GHz with respect to harmonic band rejection of the lowpass filter.

Generally speaking, this type of UWB bandpass filters can be implemented on various transmission line structures. For example, a UWB bandpass filter was reported in Reference 2 by cascading individual lowpass and highpass filters both formed on suspended strip line. Since the filter is placed in a metallic enclosure, the radiation loss from this filter is fully eliminated so as to demonstrate a low-loss UWB bandpass filter. In Reference 3, a compact UWB bandpass filter was presented on the low-temperature co-fired ceramic (LTCC) substrate by connecting an 11th-order highpass filter with a semi-lumped lowpass filter. By taking advantage of the multilayer structure, a small filter size of $0.22\lambda_g$ by $0.187\lambda_g$ by $0.0246\lambda_g$ is established.

To make up a dual wideband bandpass filter, a so-called frequency mapping approach was presented in Reference 6. The transmission line network of this filter is illustrated in Figure 7.2a. The basic idea is to create a narrow stopband in a wide passband, aiming to achieve a dual wide-passband behavior. In this aspect, a wide passband can be classified into two passbands. As shown in Figure 7.2a, the wideband virtual bandpass filter and narrowband virtual stopband filter are formed by shunt stubs with short- or open-circuited ends, and these stubs are then connected via uniform transmission lines. The lengths of the short stubs and open stubs are both equal to $\lambda_g/4$ at the center frequencies of the bandpass and bandstop filters, respectively, whereas all the connecting lines are $\lambda_g/4$ at the center frequency of the wideband bandpass filter. Since these two filters share the connecting lines, one limitation in this design is that these dual passbands cannot be spaced far apart from each other.

One pair of open- and short-circuited stubs in Figure 7.2a can be replaced by a single stepped-impedance stub, thus forming up a modified filter as shown in Figure 7.2b. These two filters may be equivalent to each other since their input admittances and slope parameters become unchanged under certain restrictions. In Reference 6, the filter in Figure 7.2b is designed on a 20-mil Rogers RO4003 substrate with dielectric constant $\varepsilon_r = 3.55$. The simulated and measured *S*-parameters

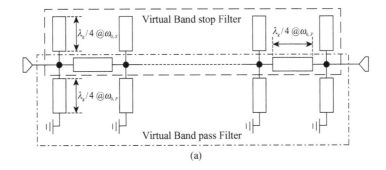

(a)

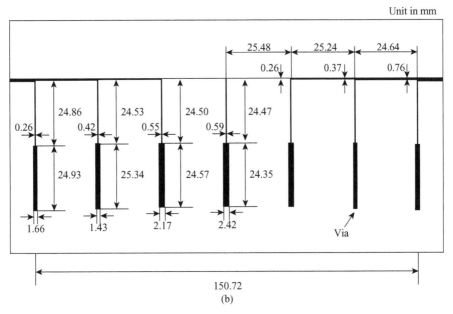

(b)

Figure 7.2. Dual wideband bandpass filter [6]. (a) Transmission line model. (b) Layout.

and group delays are depicted in Figure 7.3. The two passbands have different *FBW*s of 50.0 and 28.1%, respectively, while the *FBW* of the stopband is 19.9% under the 15-dB attenuation in measurement. The maximum group delays of the first and second bands are found 11.78 and 6.92 ns, respectively.

7.3 UWB FILTERS WITH OPTIMUM SHUNT SHORT-CIRCUITED STUBS

The basic structure of a highpass or quasi-bandpass prototype is formed by cascading shunt short-circuited stubs through uniform transmission

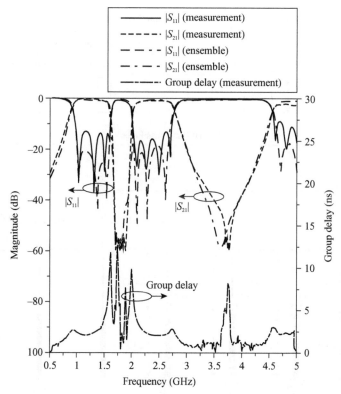

Figure 7.3. Simulated and measured results of the dual wideband bandpass filter shown in Figure 7.2b [6].

lines as discussed in Section 6.3. The word "optimum" used here means that the connecting line with length $2\theta_c$ is not redundant, and it is used to differentiate this type of filters from another type of filter, which has a quarter-wavelength in length for all the connecting lines. This structure was initially reported in Reference 7 to realize an alternative type of UWB filter. Since then, numerous works have presented a variety of UWB filters with varied geometries in References 7–11 based on the same design methodology.

7.3.1   UWB Filter with Extended Upper Stopband

Due to its periodically repeated frequency response, the basic highpass prototype has harmonic bands occurring at $\theta = 3\pi/2, 5\pi/2, \ldots$. Their existence actually degraded the upper stopband performance of the constituted UWB filter. To suppress these harmonic bands, lowpass filters

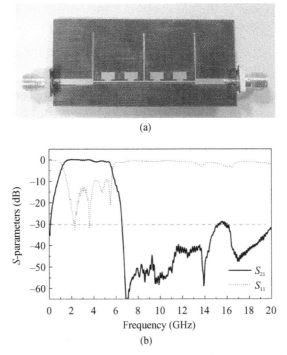

Figure 7.4. UWB filter with extended upper stopband. (a) Photograph. (b) Measured *S*-parameters [10].

can be installed externally out of the basic UWB bandpass filter. Unfortunately, it increases the overall size of the filter to a large extent. A UWB filter with internally embedded electromagnetic band-gap (EBG) structures is reported in References 10, and this filter has achieved good out-of-band performance with the help of bandgap or bandstop property in EBG section. As shown in Figure 7.4a, the EBG sections are sandwiched between two adjacent short-ended stubs instead of uniform line sections. These EBG structures are constructed as a shunt capacitively loaded transmission line, so they have the same guided-propagation properties at low frequencies belonging to the desired UWB passband. Due to the slow-wave effect of the EBG, the overall filter size is reduced from 60.6×23.65 mm$_2$ for its initial filter structure to 33.74×23.65 mm$_2$. The filter was then implemented on an Arlon substrate with the permittivity of $\varepsilon_r = 2.4$ and the thickness of $h = 0.675$ mm. As shown in Figure 7.4b, the lowest harmonic band is suppressed up to 20 GHz with the attenuation higher than 30 dB as observed in the experiment results depicted in Figure 7.4b. Moreover, the *FBW* is 140% at the center frequency of 3.4 GHz. The insertion loss and return loss in the passband are smaller than 0.9 dB and larger than 10 dB, respectively.

7.3.2 UWB Filter with Enhanced Selectivity

Instead of the above highpass or quasi-bandpass prototype oriented horizontally, twisted arrangement of its whole layout forms a modified prototype, as shown in Figure 7.5a [11]. In this aspect, the input and output ports are placed closely with each other so that a parallel-coupled line between them can be formed to create an additional signal path from the input to the output port. In other words, a cross-coupling

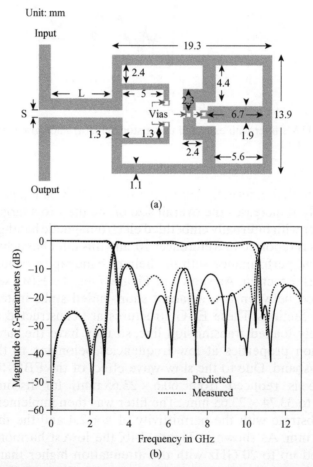

Figure 7.5. UWB filter with enhanced selectivity. (a) Layout. (b) Simulated and measured S-parameters [11].

occurs at this modified prototype. If the signal transmission in this path has the same magnitude but the opposite phase as that in the main path, two transmission zeros can be generated and allocated at pass-band edges so as to improve the filtering selectivity. The locations of these two zeros can be adjusted by S and L. A modified UWB filter was then designed and implemented on a substrate with the permittivity of 3.05 and thickness of 0.508 mm. With this twisted configuration, the filter structure occupies a reduced overall area of 13.9×26.1 mm$_2$. From the simulated and measured results shown in Figure 7.5b, we can see that the 3-dB *FWB* is 110% at the center frequency of 6.85 GHz, the mid-band insertion loss is 1.1 dB, and two pairs of transmission zeros visibly occur at passband edges.

7.4 UWB FILTERS WITH QUASI-LUMPED ELEMENTS

7.4.1 UWB Filter Implemented with Suspended Strip Line

In Reference 2, a suspended stripline UWB bandpass filter was designed based on a lumped network with series and shunt LC resonators as shown in Figure 7.6a. In this aspect, an additional capacitance is introduced in parallel to the inductance of a series resonator so as to improve the upper passband slope with emergence of transmission zero. The top and bottom views of the proposed filter are shown in Figure 7.6b, where the inductance and capacitance of each series resonator are constructed by a narrow-strip line and broadside coupling structure, respectively. Meanwhile, the inductance and capacitance of each shunt resonator are formed by a large patch that was connected to the metallic enclosure via a narrow strip conductor. Two additional capacitances are realized by the end coupling between input/output feeding lines and the patches. A substrate with a relatively high $\varepsilon_r = 10.8$ and thickness $h = 0.254$ mm is selected to attain a large capacitance. Figure 7.7a shows the photograph of the fabricated filter with the enclosure opened. From the simulated and measured results as shown in Figure 7.7b, this filter achieves the return loss higher than 10.0 dB and insertion loss in a range of 0.3–0.8 dB over the UWB passband.

7.4.2 UWB Filter Implemented on Liquid Crystal Polymer (LCP)

In Reference 12, a compact UWB filter was developed on a three-layer LCP substrate based on a lumped-circuit model, as shown in Figure 7.8a. It is composed of highpass (C_1, C_4, and L_4) and lowpass (L_1, L_2, L_3, C_2, and C_3) elements, and exhibits an UWB filtering response with

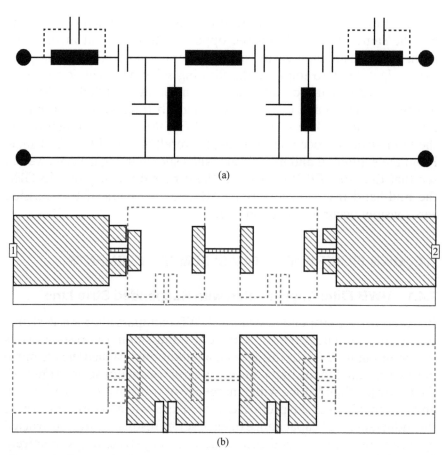

Figure 7.6. Five-pole UWB filter on suspended stripline. (a) Circuit model. (b) Top and bottom layouts [2].

six in-band transmission poles under proper assignment of these lumped elements. The layout of the filter implemented on the LCP substrate is shown in Figure 7.8b. All the quasi-lumped inductances and capacitances are constructed on the LTCC multilayer topology. The series inductances, L_1, L_2, and L_3, are realized by high-impedance microstrip line sections on the top and middle metal layers. The series capacitances, C_1 and C_4, are obtained from the broadside-coupled radial stubs between the top two mental layers, while the shunt capacitances, C_2 and C_3, are implemented by the radial stubs on the middle layer with respect to the ground plane. The shunt inductance, L_4, in the middle section of the network, is constructed by a microstrip stub with short-circuited end on the top metal layer.

The LCP substrate has a dielectric constant $\varepsilon_r = 3.15$ and loss tangent $\tan\delta = 0.0025$. The thicknesses of two individual layers are $h_1 = 0.1$ and

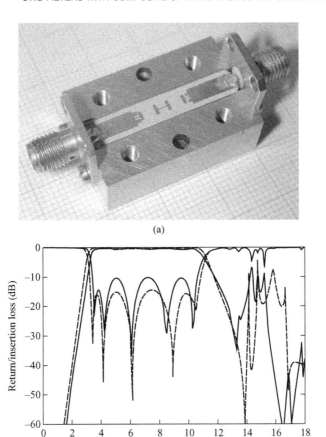

(a)

(b)

Figure 7.7. Suspended stripline UWB filter. (a) Photograph. (b) Simulated and measured S-parameters [2].

$h_2 = 0.7$ mm, respectively. The simulated and measured frequency responses are plotted in Figure 7.9. The measured 3-dB *FBW* is found 95.7% at the center frequency of 6.9 GHz. The minimum insertion loss occurs at 6.16 GHz, which is 0.35 dB, and the return loss in the passband is larger than 10 dB. Moreover, this filter attains a good upper stopband performance with 28.1 dB suppression up to 18 GHz.

7.5 UWB FILTERS WITH COMPOSITE CPW AND MICROSTRIP STRUCTURE

In Reference 13, a UWB bandpass filter is presented based on microstrip line and CPW placed on the top and bottom interfaces of a substrate

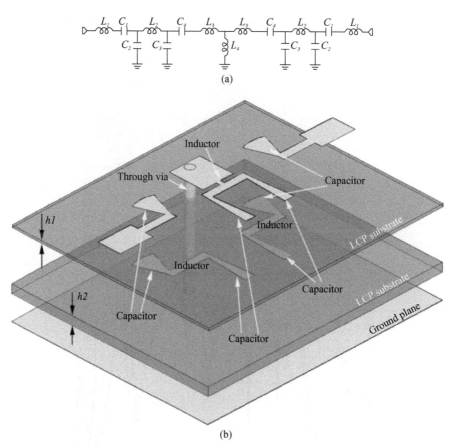

Figure 7.8. Six-pole UWB filter using LCP technology. (a) Circuit model. (b) 3D schematic layout [12].

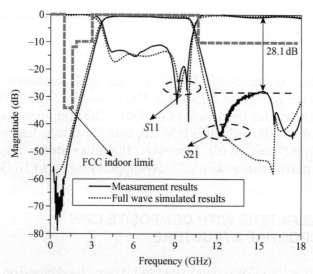

Figure 7.9. Simulated and measured *S*-parameter of the UWB filter shown in Figure 7.8b [12].

as shown in Figure 7.10a, where a broadside coupled line is utilized to get a high coupling degree. Instead of a half-wavelength CPW resonator as usually used, the UWB filter herein forms a quarter-wavelength CPW on the ground plane and microstrip-to-CPW broadside coupling on the top metal plane. Alternatively, the quarter-wavelength CPW resonator can be considered as a half-wavelength slotline resonator on the ground plane. In addition, two feed lines are arranged in parallel to each other, and a cross coupling path is introduced to generate two transmission zeros at the passband edges to sharpen the rejection skirts. Following the early discussion, we can understand that the locations of these transmission zeros are controlled by the coupling gap in a parallel-coupled line. Next, this filter was implemented on a GML1000 substrate with a permittivity of 3.05 and a thickness of 0.508 mm. The measured 3-dB *FBW* is 90% at the center frequency of 6.4 GHz as depicted in Figure 7.10b. The mid-band insertion loss is 0.6 dB, and the return loss is larger than 20 dB. Two transmission zeros are observed at 1.95 and 10.36 GHz, respectively. By fully utilizing the two metal layers, this UWB filter is quite compact and only has an overall size of $0.25\lambda_g \times 0.08\lambda_g$ at the center frequency.

A similar UWB bandpass filter with composite CPW and microstrip line was proposed in Reference 14. Different from the analysis method in Reference 13, the UWB filter in Reference 14 is designed based on a lumped element model as discussed in Section 7.4, and they are realized by the CPW structure on the bottom plane and the feeding microstrip lines on the top metal. The schematic layout of this filter is shown in Figure 7.11. With this two-layer orientation, a miniaturized filter with a size of $0.306\lambda_g \times 0.458\lambda_g$ is achieved. Looking at Figure 7.11, this UWB filter can also be considered to be composed of a quarter-wavelength CPW resonator and microstrip-to-CPW coupling structure.

7.6 UWB FILTER WITH MICROSTRIP RING RESONATOR

The concept of utilizing a ring resonator to design bandpass filter with a wide passband was initially proposed in Reference 15. Two open-circuited stubs are attached with a ring resonator to create a wide passband relying on the lower and upper transmission zeros. Later on, a bandpass filter with two ring resonators reportedly achieves an *FWB* of 80% in Reference 16. Both works used direct-connected or tapped feeding structures at two ports, thereby suffering from poor rejection performance in a frequency range below the desired passband. A more

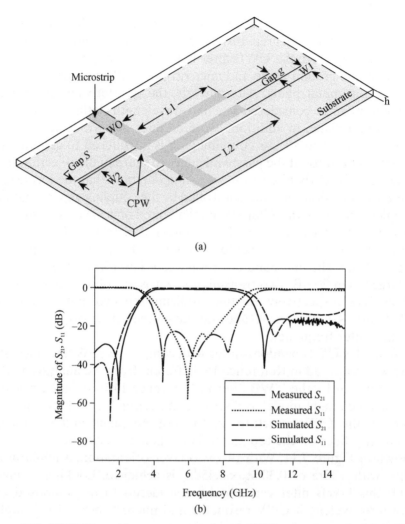

Figure 7.10. UWB filter with a quarter-wavelength CPW resonator and microstrip to CPW feeding structure [13]. (a) Layout with the dimensions: $W_0 = 1.2$ mm, $W_1 = 0.8$ mm, $L_1 = 6.2$ mm, $W_2 = 2.4$ mm, $L_2 = 8.6$ mm, $s = 0.2$ mm. (b) Simulated and measured S-parameter.

generalized wideband ring resonator bandpass filter was reported in Reference 17 as shown in Figure 7.12. Due to noncontact capacitive feeding at two ports, this ring resonator filter can block signal transmission at lower frequencies and improve the out-of-band performance.

The wideband filter in Figure 7.12 is composed of a quintuple-mode resonator and interdigital coupling structure at input and output. Even-

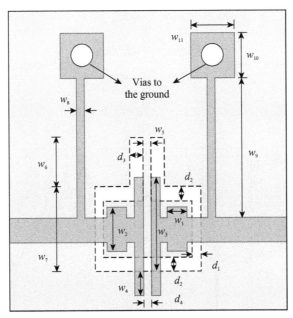

Figure 7.11. Schematic layout of an UWB filter with a composite CPW and microstrip line structure [14].

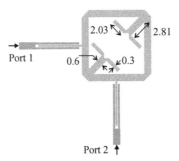

Figure 7.12. Schematic layout of a wideband quintuple-mode ring resonator bandpass with internally embedded T-shaped stubs [17]. All units are in mm.

and odd-mode analysis can be applied to quintuple-mode ring resonator with a symmetrical plane in the diagonal direction. The five resonant modes can be simultaneously excited and their resonant frequencies can be determined under the transverse resonant condition that the sum of two oppositely oriented input impedances at the same position be zero. Figure 7.13 shows the equivalent circuit of this quintuple-mode

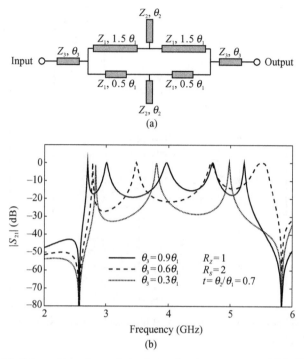

Figure 7.13. (a) Schematics of a quintuple-mode ring resonator [17]. (b) Frequency responses of S_{21}-magnetudes under varied orthogonal stub lengths (θ_3) under weak coupling at two ports.

ring resonator and the graphs of its resonant frequencies with respect to the two stubs at two ports. If the length of two identical stubs $\theta_3 = 0$ or are very small, the ring resonator is in fact a triple-mode resonator. If the two open stubs at the corner of the square ring are stretched close to one-eighth of a guided-wavelength, the first two even-order resonant frequencies can be moved down and symmetrically located at the two sides of its first odd-order counterpart, as shown in Figure 7.13b. When the length of the extra stubs at two port increases, two additional coupling peaks are moved into the desired passband, thus causing synchronous appearance of five or quintuple resonances, as shown in Figure 7.13b.

With the use of the interdigital coupling structure with enhanced degree, a wide passband can be constructed as well. With proper adjustment of these coupling structures, the upper stopband of the constituted filter can be reasonably widened with high attenuation. This filter is then designed and implemented on a substrate with a permittivity of

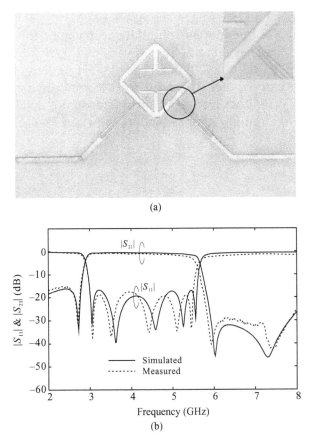

(a)

(b)

Figure 7.14. Quintuple-resonance ring resonator bandpass filter. (a) Photograph. (b) Simulated and measured S-parameters [17].

10.8 and a thickness of 0.635 mm. Figure 7.14a is the photograph of the fabricated filter. The simulated and measured frequency responses are plotted in Figure 7.14b. The measured 3-dB *FBW* is 64% at the center frequency of 4.25 GHz. The return loss is lower than 17.5 dB over the passband, and the insertion loss at center is about 0.9 dB.

7.7 UWB FILTER USING MULTILAYER STRUCTURES

On the one hand, a strong coupling structure or large capacitance is always required in design of a bandpass filter with an ultra-wide bandwidth. In this context, multilayer structure allows one to achieve this target by taking advantage of the broadside-coupling between layers as reported [18–24]. On the other hand, by making use of the vertical degree of freedom in a multilayout structure, the filter size can be reduced to a large extent. So far, various UWB filters are implemented on LCP [5, 12, 18–20] or LTCC [21–23], and all of them can be placed in this category regardless of varied complicated configurations. Of them, a UWB filter on LTCC using a T-resonator will be specifically discussed below, and its circuit model is shown in Figure 7.15a.

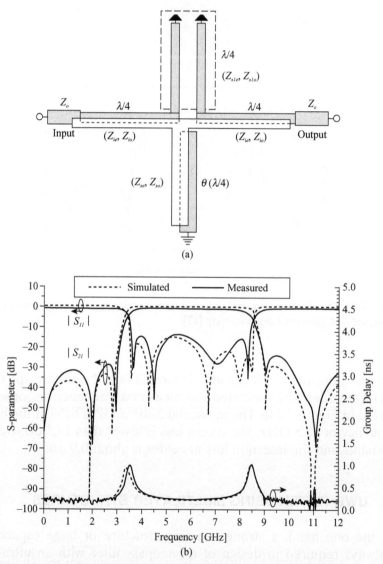

Figure 7.15. UWB filter with T-resonator on LTCC [22]. (a) Transmission line network. (b) Simulated and measured S-parameters.

This filter is composed of a T-resonator with a $\lambda_g/4$ open-ended stub and a capacitively loaded a $\lambda_g/4$ short-circuited stub in the broad-side coupling section. This arrangement aims to generate two transmission zeros so as to specify the two desired passband edges. In the meantime, the capacitive coupling structure at two ports effectively blocks the DC signal. Two types of feeding lines can be used, that is, $\lambda_g/4$ line with open end and $\lambda_g/2$ line with short end, as shown in Figure 7.15a. In design, the effective coupling length is kept as $\lambda_g/4$, and the rest is bended as depicted in the dashed boxes of the figure. This arrangement introduces a cross-coupling path so as to create two transmission zeros near the lower and upper stopbands. Compared with the early-reported ones, this filter has the improved out-of-band performance. Figure 7.15b exhibits an elliptic filtering response, where the measured 3-dB *FBW* is 86% at the center frequency of 5.95 GHz, the insertion loss is smaller than 0.9 dB, and the return loss is better than 14 dB in the whole pass-band. The maximum group delay in a range from 3.8 to 8.2 GHz is about 0.2 ns.

7.8   UWB FILTER WITH SUBSTRATE INTEGRATED WAVEGUIDE (SIW)

Besides the planar integrated UWB filters we discussed so far, the UWB filters can also be implemented on a substrate integrated wave-guide (SIW) [25, 26]. The SIW [27] or the laminated waveguide [28] primarily consists of a dielectric substrate sandwiched by two metal plates on the top and bottom, which are electrically connected by vias at two side edges. As the inherent property of a waveguide, this SIW itself has much higher Q-factor than its planar transmission line struc-ture, and is believed useful in design of planar filters with a relatively narrow bandwidth. A seven-pole SIW UWB filter was developed [26], and its schematic layout is shown in Figure 7.16a. This filter is composed of five SIW resonators that cascade in a zigzag manner. Its input and output SIW sections are connected to the main feeding line through the SIW-to-microstrip transitions. With this arrangement, the filter becomes much compact than those with inline configuration. More importantly, cross-coupling is introduced between the nonadjacent resonators, thus its out-of band performance can be improved. The cross-coupling degree can be controlled by removing one or two vias from the decoupling walls with the width of dw_i ($i = 1, 2, 3$), as shown in Figure 7.16a.

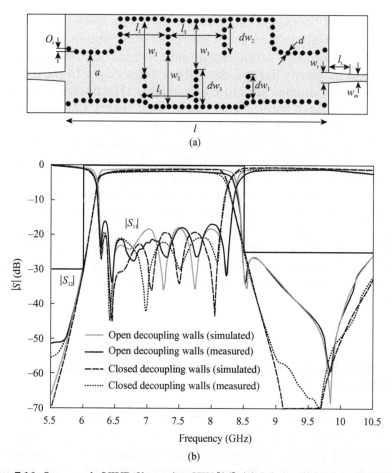

(a)

(b)

Figure 7.16. Seven-pole UWB filter using SIW [26]. (a) Schematic layout. Dimensions: $o_i = 0.9$ mm, $a = 12.80$ mm, $l = 76.62$ mm, $l_1 = 12.74$ mm, $l_2 = 14.11$ mm, $l_3 = 15.40$ mm, $w_1 = 17.23$ mm, $w_2 = 14.43$ mm, $w_3 = 15.64$ mm, $dw_1 = 7.13$ mm, $dw_2 = 9.35$ mm, $dw_3 = 7.76$ mm, $w_t = 2.8$ mm, $l_t = 7.0$ mm, $w_m = 1.85$ mm and $d = 1.05$ mm. (b) Synthesized and simulated S-parameters.

The seven-pole UWB filter was designed on a Rogers RO4003C substrate with dielectric constant $\varepsilon_r = 3.60$, thickness $h = 0.813$, and loss tangent $\tan\delta = 0.0027$. The simulated and measured results with and without removed vias in the decoupling walls are demonstrated in Figure 7.16b. The filter is designed targeting the European UWB radiation mask with 30 and 25 dB rejection at 6.0 and 8.5 GHz, respectively. The measured results show that the filter with cross-coupling fits the emission mask better. It achieves a return loss larger than 17.0 dB and an insertion loss smaller than 1.18 dB, excluding the connector loss in measurement.

7.9 UWB FILTER WITH NOTCH BAND

As known, the power density for UWB system is very low and it does not cause any interference to the existing radio services. However, the latter one with relatively high power may seriously disturb the communication for the newly developed UWB systems. In order to eliminate or reduce those interferences, UWB filters with single or multiple notch bands were studied and explored as reported in References 29–38. Most of the notch band UWB filters is realized by designing an initial UWB filter and then introducing notch band elements. So far, there are three ways to introduce notch bands in the UWB main passband. Besides the asymmetrical coupling structures at the input and output ports discussed in Chapter 5, the notch band UWB filter can be designed using open stubs [29–32] and notching resonators [33–38].

A notch band UWB filter [29] was designed based on the UWB filter presented in Reference [11]. The schematic layout of this filter is shown in Figure 7.17a. The notch band is generated by replacing the connecting line by a stub-embedded line between adjacent short-circuited stubs. Compared with the conventional open-ended stub on the transmission line with a T-junction and spur line, the embedded stub configuration provides a narrow rejection band or called "notch band." The length of the embedded stub is $\lambda_g/4$ at the notching frequency. Hence, the notch band location can be adjusted with the length of this stub. One prototype filter is implemented on a substrate with a permittivity of 3.05 and a thickness of 0.508 mm. The simulated and measured S-parameters and group delays are depicted in Figure 7.17b. The measured 3-dB bandwidth of the UWB filter is 110% at the center frequency of 6.85 GHz. A notch band occurs at 5.83 GHz with 10-dB *FBW* of 4.6%. The attenuation at the notch band is larger than 23 dB. The group delay is about 0.5 ns at the mid-band frequency of the UWB passband.

The other common way to introduce one or multiple notch bands is to weakly couple a resonator with its resonant frequency at the notch band to the main UWB filter, as shown in Figure 7.18a. A folded SIR is chosen as the notching resonator due to its compact size and higher harmonic resonance. As discussed in Reference [33], the

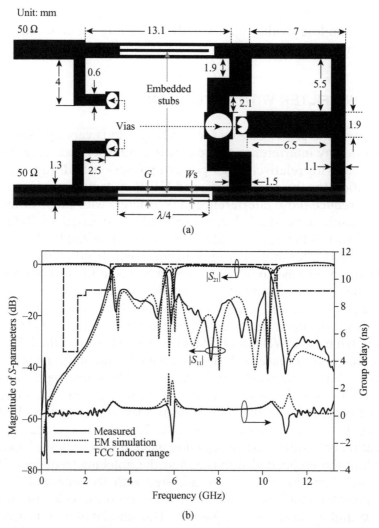

Figure 7.17. Notch band UWB filter with embedded open-circuited stub [29]. (a) Schematic layout. (b) Simulated and measured S-parameters.

notching resonators are placed on the third metal layer underneath the UWB filters formed on the top and second metal layers in the LCP substrate. The coupling strength can be controlled by a thickness of the layer between the second and third metal layers, and it is demonstrated that the thicker that substrate, the narrower the notch band will be. The location of the notch band is determined by the overall length of the folded resonator, and multiple notch bands can be produced by using more than one notching resonators with different lengths.

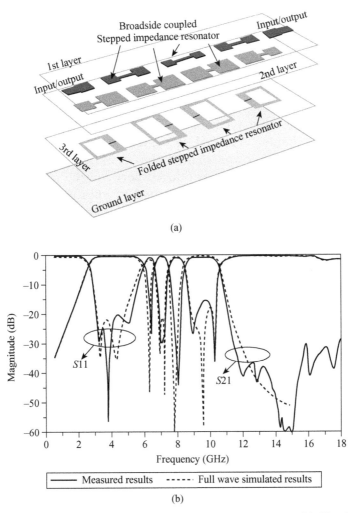

(a)

(b)

Figure 7.18. Notch band UWB filter [33]. (a) Schematic layout. (b) Simulated and measured S-parameters.

The UWB filter with two notch bands was designed. The LCP substrate has a dielectric constant $\varepsilon_r = 3.15$ and loss tangent $\tan\delta = 0.0025$. The substrate thicknesses between the first and the second, the second and the third, the third and the ground plane are 50, 450, and 300 µm, respectively. The simulated and measured S-parameters are shown in Figure 7.18b. Two notch bands in the main UWB passband are observed. Their measured 3-dB FBWs are 9.5 and 13.4% at the center frequencies of 6.4 and 8 GHz, respectively. The notch band attenuations are 26.4 and 43.7 dB. The UWB passband is actually divided into three passbands with *FBW*s of 78.5, 11.5, and 21.2% at their own center frequencies of 4.28, 7.08, and 9.53 GHz, respectively.

7.10   SUMMARY

A review of the other types of UWB bandpass filters has been presented in this chapter. A basic UWB filter can be obtained by combining a few filters, such as lowpass-and-highpass, and lowpass-and-bandpass. As a cross coupling path is introduced between the input and output ports, transmission zeros can be created to improve the out-of-band performance. Compared with the single-layer structure, the multilayer structure provides more flexibility in constructing coupling structure and making the overall size smaller. But it may bring out the complexity in design for multiple metal layers. In order to avoid interference from existing radio systems, several UWB filters with single or multiple notch bands were reviewed at last.

REFERENCES

1   C. L. Hsu, F. C. Hsu, and J. T. Kuo, "Microstrip bandpass filter for ultra-wideband (UWB) wireless communications," *IEEE MTT-S Int. Microwave Symp. Dig.*, 2005, pp. 679–682.

2   W. Menzel, M. S. Rahman, and L. Zhu, "Low-loss ultra-wideband (UWB) filters using suspended stripline," *Asia-Pacific Microwave Conf. Proc.*, APMC 2005, vol. 4, December 2005, pp. 2148–2151.

3   C. H. Wu, Y. S. Lin, C. H. Wang, and C. H. Chen, "A compact LTCC ultra-wideband bandpass filter using semi-lumped parallel resonance circuits for spurious suppression," *Proc. 37th European Microwave Conf.*, October 2007, pp. 532–535.

4   C.-W. Tang and M. G. Chen, "A microstrip ultra-wideband bandpass filter with cascaded broadband bandpass and bandstop filters," *IEEE Trans. Microw. Theory Tech.* 55 (2007) 2412–2418.

5   Z.-C. Hao and J.-S. Hong, "UWB bandpass filter using cascaded miniature high-pass and low-pass filters with multilayer liquid crystal polymer technology," *IEEE Trans. Microw. Theory Tech.* 58(4) (2010) 941–948.

6   A.-S. Liu, T.-Y. Huang, and R.-B. Wu, "A dual wideband filter design using frequency mapping and stepped—impedance resonators," *IEEE Trans. Microw. Theory Tech.* 56(12) (2008) 2921–2929.

7   J.-S. Hong and H. Shaman, "An optimum ultra-wideband microstrip filter," *Microw. Optical Tech. Lett.* 47(3) (2005) 230–233.

8 W. T. Wong, Y. S. Lin, C. H. Wang, and C. H. Chen, "Highly selective microstrip bandpass filters for ultra-wideband applications," *Proc. Asia-Pacific Microwave Conf.*, November 2005, pp. 2850–2853.

9 P. Cai, Z. Ma, X. Guan, X. Yang, Y. Kobayashi, T. Anada, and G. Hagiwara, "A compact UWB bandpass filter using two section opencircuited stubs to realize transmission zeros," *Asia-Pacific Microw. Conf.*, vol. 5, December 2005, p. 14.

10 J. Garcia-Garcia, J. Bonache, and F. Martin, "Application of electromagnetic bandgaps to the design of ultra-wide bandpass filters with good out-of-band performance," *IEEE Trans. Microw. Theory Tech.* 54 (2006) 4136–4140.

11 H. Shaman and J.-S. Hong, "A novel ultra-wideband (UWB) bandpass filter (BPF) with pairs of transmission zeroes," *IEEE Microw. Wireless Compon. Lett.* 17(2) (2007) 121–123.

12 Z.-C. Hao and J.-S. Hong, "Ultra-wideband bandpass filter using multilayer liquid-crystal-polymer technology," *IEEE Trans. Microw. Theory Tech.* 56(9) (2008) 2095–2100.

13 N. Thomson and J.-S. Hong, "Compact ultra-wideband microstrip/coplanar waveguide bandpass filter," *IEEE Microw. Wireless Compon. Lett.* 17(3) (2007) 184–186.

14 T. N. Kuo, S. C. Lin, and C. H. Chen, "Compact ultra-wideband bandpass filter using composite microstrip-coplanar-waveguide structure," *IEEE Trans. Microw. Theory Tech.* 54 (2006) 3772–3778.

15 L. H. Hsieh and K. Chang, "Compact, low insertion-loss, sharp-rejection, and wide-band microstrip bandpass filters," *IEEE Trans. Microw. Theory Tech.* 51(4) (2003) 1241–1246.

16 W. Liu, Z. Ma, C. P. Chen, G. Zheng, and T. Anada, "A novel UWB filter using a new type of microstrip double-ring resonators," *Asia-Pacific Microwave Conf.*, December 2006, pp. 33–36.

17 S. Sun and L. Zhu, "Wideband microstrip ring resonator bandpass filters under multiple resonances," *IEEE Trans Microw. Theory Tech.* 55(10) (2007) 2176–2182.

18 Z.-C. Hao and J.-S. Hong, "Ultra wideband bandpass filter using embedded stepped impedance resonators on multilayer liquid crystal polymer substrate," *IEEE Microw. Wireless Compon. Lett.* 18(9) (2008) 581–583.

19 Z.-C. Hao and J.-S. Hong, "Compact wide stopband ultra wideband bandpass filter using multilayer liquid crystal polymer technology," *IEEE Microw. Wireless Compon. Lett.* 19(5) (2009) 290–292.

20 Z.-C. Hao and J.-S. Hong, "Quasi-elliptic UWB bandpass filter using multilayer liquid crystal polymer technology," *IEEE Microw. Wireless Compon. Lett.* 20(4) (2010) 202–204.

21 A. M. Abbosh, "Planar bandpass filters for ultra-wideband applications," *IEEE Trans Microw. Theory Tech.* 55(10) (2007) 2262–2269.

22 T. H. Duong and I. S. Kim, "New elliptic function type UWB BPF based on capacitively coupled 1/4 open T resonator," *IEEE Trans. Microw. Theory Tech.* 57 (2009) 3089–3098.

23 J. A. Ruiz-Cruz, Y. Zhang, K. A. Zaki, A. J. Piloto, and J. Tallo, "Ultra-wideband LTCC ridge waveguide filters," *IEEE Microw. Wireless Compon. Lett.* 17(2) (2007) 115–117.

24 T. H. Duong and I. S. Kim, "Steeply sloped UWB bandpass filter based on stub-loaded resonator," *IEEE Microw. Wireless Compon. Lett.* 20(8) (2010) 441–443.

25 Z. Hao, W. Hong, J. Chen, X. Chen, and K., "Compact super-wide bandpass substrate integrated waveguide (SIW) filters," *IEEE Trans. Microw. Theory Tech.* 53(9) (2005) 2968–2977.

26 F. Mira, J. Mateu, S. Cogollos, and V. E. B'oria, "Design of ultra-wideband substrate integrated waveguide filters in zigzag topology," *IEEE Microw. Wireless Compon. Lett.* 19(5) (2009) 281–283.

27 D. Deslandes and K. Wu, "Integrated microstrip and rectangular waveguide in planar form," *IEEE Microw. Wireless Comp. Lett.* 11 (2001) 68–70.

28 H. Uchimura and T. Takenoshita, "Development of a laminated waveguide," *IEEE Trans. Microw. Theory Tech.* 46 (1998) 2438–2443.

29 H. Shaman and J.-S. Hong, "Ultra-wideband (UWB) bandpass filter with embedded band notch structures," *IEEE Microw. Wireless Compon. Lett.* 17(3) (2007) 193–195.

30 G.-M. Yang, R. Jin, C. Vittoria, V. G. Harris, and N. X. Sun, "Small ultra-wideband (UWB) bandpass filter with notch band," *IEEE Microw. Wireless Compon. Lett.* 18(3) (2008) 176–178.

31 Y.-H. Chun, H. Shaman, and J.-S. Hong, "Switchable embedded notch structure for UWB bandpass filter," *IEEE Microw. Wireless Compon. Lett.* 18(9) (2008) 590–592.

32 M.-H. Weng, C.-T. Liauh, H.-W. Wu, and S. R. Vargas, "An ultra-wideband bandpass filter with an embedded open-circuited stub structure to improve in-band performance," *IEEE Microw. Wireless Compon. Lett.* 19(3) (2009) 146–148.

33 Z.-C. Hao and J.-S. Hong, "Compact UWB filter with double notch-bands using multilayer LCP technology," *IEEE Microw. Wireless Compon. Lett.* 19(8) (2009) 500–502.

34 S.-G. Mao, Y.-Z. Chueh, C.-H. Chen, and M.-C. Hsieh, "Compact ultra-wideband conductor-backed coplanar waveguide bandpass filter with a dual band-notched response," *IEEE Microw. Wireless Compon. Lett.* 19(3) (2009) 149–151.

35 Z.-C. Hao, J.-S. Hong, J. P. Parry, and D. P. Hand, "Ultra-wideband bandpass filter with multiple notch band using nonuniform periodical slotted

ground structure," *IEEE Trans. Microw. Theory Tech.* 57(12) (2009) 3080–3088.

36 X. Luo, J.-G. Ma, K. Ma, and K. S. Yeo, "Compact UWB bandpass filter with ultra narrow notched band," *IEEE Microw. Wireless Compon. Lett.* 20(3) (2009) 145–147.

37 W.-J. Lin, J.-Y. Li, L.-S. Chen, D.-B. Lin, and M.-P. Houng, "Investigation in open circuited metal line embedded in defected ground structure and its applications to UWB filters," *IEEE Microw. Wireless Compon. Lett.* 20(3) (2010) 148–150.

38 S. Pirani, J. Nourinia, and C. Ghobadi, "Band-notched UWB BPF design using parasitic coupled line," *IEEE Microw. Wireless Compon. Lett.* 20(8) (2010) 444–446.

INDEX

Microwave Bandpass Filters for Wideband Communications, First Edition. Lei Zhu, Sheng Sun, Rui Li.
© 2012 John Wiley & Sons, Inc. Published 2012 by John Wiley & Sons, Inc.

WILEY SERIES IN MICROWAVE AND OPTICAL ENGINEERING

KAI CHANG, Editor
Texas A&M University

MICROSTRIP CIRCUITS • *Fred Gardiol*

HIGH-SPEED VLSI INTERCONNECTIONS, second Edition • *Ashok K. Goel*

FUNDAMENTALS OF WAVELETS: THEORY, ALGORITHMS, AND APPLICATIONS, Second Edition • *Jaideva C. Goswami and Andrew K. Chan*

HIGH-FREQUENCY ANALOG INTEGRATED CIRCUIT DESIGN • *Ravender Goyal (ed.)*

RF AND MICROWAVE TRANSMITTER DESIGN • *Andrei Grebennikov*

ANALYSIS AND DESIGN OF INTEGRATED CIRCUIT ANTENNA MODULES • *K. C. Gupta and Peter S. Hall*

PHASED ARRAY ANTENNAS, second Edition • *R. C. Hansen*

STRIPLINE CIRCULATORS • *Joseph Helszajn*

THE STRIPLINE CIRCULATOR: THEORY AND PRACTICE • *Joseph Helszajn*

LOCALIZED WAVES • *Hugo E. Hernández-Figueroa, Michel Zamboni-Rached, and Erasmo Recami (eds.)*

MICROSTRIP FILTERS FOR RF/MICROWAVE APPLICATIONS, second Edition • *Jia-Sheng Hong*

MICROWAVE APPROACH TO HIGHLY IRREGULAR FIBER OPTICS • *Huang Hung-Chia*

NONLINEAR OPTICAL COMMUNICATION NETWORKS • *Eugenio Iannone, Francesco Matera, Antonio Mecozzi, and Marina Settembre*

FINITE ELEMENT SOFTWARE FOR MICROWAVE ENGINEERING • *Tatsuo Itoh, Giuseppe Pelosi, and Peter P. Silvester (eds.)*

INFRARED TECHNOLOGY: APPLICATIONS TO ELECTROOPTICS, PHOTONIC DEVICES, AND SENSORS • *A. R. Jha*

SUPERCONDUCTOR TECHNOLOGY: APPLICATIONS TO MICROWAVE, ELECTRO-OPTICS, ELECTRICAL MACHINES, AND PROPULSION SYSTEMS • *A. R. Jha*

TIME AND FREQUENCY DOMAIN SOLUTIONS OF EM PROBLEMS USING INTEGRAL EQUATIONS AND A HYBRID METHODOLOGY • *B. H. Jung, T. K. Sarkar, S. W. Ting, Y. Zhang, Z. Mei, Z. Ji, M. Yuan, A. De, M. Salazar-Palma, and S. M. Rao*

OPTICAL COMPUTING: AN INTRODUCTION • *M. A. Karim and A. S. S. Awwal*

INTRODUCTION TO ELECTROMAGNETIC AND MICROWAVE ENGINEERING • *Paul R. Karmel, Gabriel D. Colef, and Raymond L. Camisa*

MILLIMETER WAVE OPTICAL DIELECTRIC INTEGRATED GUIDES AND CIRCUITS • *Shiban K. Koul*

ADVANCED INTEGRATED COMMUNICATION MICROSYSTEMS • *Joy Laskar, Sudipto Chakraborty, Manos Tentzeris, Franklin Bien, and Anh-Vu Pham*

MICROWAVE DEVICES, CIRCUITS AND THEIR INTERACTION • *Charles A. Lee and G. Conrad Dalman*

ADVANCES IN MICROSTRIP AND PRINTED ANTENNAS • *Kai-Fong Lee and Wei Chen (eds.)*

SPHEROIDAL WAVE FUNCTIONS IN ELECTROMAGNETIC THEORY • *Le-Wei Li, Xiao-Kang Kang, and Mook-Seng Leong*

ARITHMETIC AND LOGIC IN COMPUTER SYSTEMS • *Mi Lu*

OPTICAL FILTER DESIGN AND ANALYSIS: A SIGNAL PROCESSING APPROACH • *Christi K. Madsen and Jian H. Zhao*

THEORY AND PRACTICE OF INFRARED TECHNOLOGY FOR NONDESTRUCTIVE TESTING • *Xavier P. V. Maldague*

METAMATERIALS WITH NEGATIVE PARAMETERS: THEORY, DESIGN, AND MICRO-WAVE APPLICATIONS • *Ricardo Marqués, Ferran Martín, and Mario Sorolla*

OPTOELECTRONIC PACKAGING • *A. R. Mickelson, N. R. Basavanhally, and Y. C. Lee (eds.)*

FUNDAMENTALS OF GLOBAL POSITIONING SYSTEM RECEIVERS: A SOFTWARE APPROACH, second Edition • *James Bao-yen Tsui*

SUBSURFACE SENSING • *Ahmet S. Turk, A. Koksal Hocaoglu, and Alexey A. Vertiy (eds.)*

RF/MICROWAVE INTERACTION WITH BIOLOGICAL TISSUES • *André Vander Vorst, Arye Rosen, and Youji Kotsuka*

InP-BASED MATERIALS AND DEVICES: PHYSICS AND TECHNOLOGY • *Osamu Wada and Hideki Hasegawa (eds.)*

COMPACT AND BROADBAND MICROSTRIP ANTENNAS • *Kin-Lu Wong*

DESIGN OF NONPLANAR MICROSTRIP ANTENNAS AND TRANSMISSION LINES • *Kin-Lu Wong*

PLANAR ANTENNAS FOR WIRELESS COMMUNICATIONS • *Kin-Lu Wong*

FREQUENCY SELECTIVE SURFACE AND GRID ARRAY • *T. K. Wu (ed.)*

ACTIVE AND QUASI-OPTICAL ARRAYS FOR SOLID-STATE POWER COMBINING • *Robert A. York and Zoya B. Popovic (eds.)*

OPTICAL SIGNAL PROCESSING, COMPUTING AND NEURAL NETWORKS • *Francis T. S. Yu and Suganda Jutamulia*

ELECTROMAGNETIC SIMULATION TECHNIQUES BASED ON THE FDTD METHOD • *Wenhua Yu, Xiaoling Yang, Yongjun Liu, and Raj Mittra*

SiGe, GaAs, AND InP HETEROJUNCTION BIPOLAR TRANSISTORS • *Jiann Yuan*

PARALLEL SOLUTION OF INTEGRAL EQUATION-BASED EM PROBLEMS • *Yu Zhang and Tapan K. Sarkar*

ELECTRODYNAMICS OF SOLIDS AND MICROWAVE SUPERCONDUCTIVITY • *Shu-Ang Zhou*

MICROWAVE BANDPASS FILTERS FOR WIDEBAND COMMUNICATIONS • *Lei Zhu, Sheng Sun, and Rui Li*

Lightning Source UK Ltd.
Milton Keynes UK
UKHW041529270219
338058UK00003B/114/P